About the Author

LUDWIG WITTGENSTEIN (1889–1951) was born in Austria and studied at Cambridge under Bertrand Russell. Having volunteered to serve in the Austrian army at the outbreak of World War I, he was captured in 1918 and sent to a prison camp in Italy, where he finished his only published work, *Tractatus Logico-Philosophicus*, one of the most important philosophical works of all time. After the war, Wittgenstein eventually returned to Cambridge to teach. His posthumous publications include *Philosophical Investigations*, *The Blue and Brown Books*, *On Certainty*, and several volumes of notebooks and lectures.

MAJOR
WORKS

MAJOR WORKS

SELECTED
PHILOSOPHICAL
WRITINGS

LUDWIG WITTGENSTEIN

HARPER**PERENNIAL** **x** MODERN**THOUGHT**

NEW YORK • LONDON • TORONTO • SYDNEY • NEW DELHI • AUCKLAND

HARPER**PERENNIAL** ⊗ MODERN**THOUGHT**

HarperCollins books may be purchased for educational, business, or
sales promotional use. For information, please e-mail the Special Mar-
kets Department at SPsales@harpercollins.com.

FIRST EDITION

Designed by Laura Kaeppel

Library of Congress Cataloging-in-Publication Data is available upon
request.

ISBN 978-0-06-155024-9

15 16 17 18   /RRD   10 9

TRACTATUS
LOGICO-PHILOSOPHICUS

Translated by C. K. Ogden

PREFACE

THIS BOOK WILL PERHAPS ONLY BE UNDERSTOOD BY THOSE who have themselves already thought the thoughts which are expressed in it—or similar thoughts. It is therefore not a text-book. Its object would be attained if there were one person who read it with understanding and to whom it afforded pleasure.

The book deals with the problems of philosophy and shows, as I believe, that the method of formulating these problems rests on the misunderstanding of the logic of our language. Its whole meaning could be summed up somewhat as follows: What can be said at all can be said clearly; and whereof one cannot speak thereof one must be silent.

The book will, therefore, draw a limit to thinking, or rather—not to thinking, but to the expression of thoughts; for, in order to draw a limit to thinking we should have to be able to think both sides of this limit (we should therefore have to be able to think what cannot be thought).

The limit can, therefore, only be drawn in language and what lies on the other side of the limit will be simply nonsense.

How far my efforts agree with those of other philosophers I will not decide. Indeed what I have here written makes no claim to novelty in points of detail; and therefore I give no sources, because it is indifferent to me whether what I have thought has already been thought before me by another.

I will only mention that to the great works of Frege and the writings of my friend Bertrand Russell I owe in large measure the stimulation of my thoughts.

If this work has a value it consists in two things. First that in it thoughts are expressed, and this value will be the greater the better the thoughts are expressed. The more the nail has been hit on the head.—Here I am conscious that I have fallen far short of the possible. Simply because my powers are insufficient to cope with the task.—May others come and do it better.

On the other hand the *truth* of the thoughts communicated here seems to me unassailable and definitive. I am, therefore, of the opinion that the problems have in essentials been finally solved. And if I am not mistaken in this, then the value of this work secondly consists in the fact that it shows how little has been done when these problems have been solved.

LUDWIG WITTGENSTEIN
Vienna, 1918

TRACTATUS LOGICO-PHILOSOPHICUS

1*   The world is everything that is the case.

1.1   The world is the totality of facts, not of things.

1.11   The world is determined by the facts, and by these be-
        ing *all* the facts.

1.12   For the totality of facts determines both what is the
        case, and also all that is not the case.

1.13   The facts in logical space are the world.

1.2   The world divides into facts.

1.21   Anyone can either be the case or not be the case, and
        everything else remain the same.

2   What is the case, the fact, is the existence of atomic facts.

---

* The decimal figures as numbers of the seperate propositions indicate the logical
importance of the propositions, the emphasis laid upon them in my exposition. The
propositions *n*.1, *n*.2, *n*.3, etc., are comments on proposition No. *n*; the propositions *n.m*1, *n.m*2, etc., are comments on the proposition No. *n.m*; and so on.

2.01   An atomic fact is a combination of objects (entities, things).

2.011   It is essential to a thing that it can be a constituent part of an atomic fact.

2.012   In logic nothing is accidental: if a thing *can* occur in an atomic fact the possibility of that atomic fact must already be prejudged in the thing.

2.0121   It would, so to speak, appear as an accident, when to a thing that could exist alone on its own account, subsequently a state of affairs could be made to fit.

     If things can occur in atomic facts, this possibility must already lie in them.

     (A logical entity cannot be merely possible. Logic treats of every possibility, and all possibilities are its facts.)

     Just as we cannot think of spatial objects at all apart from space, or temporal objects apart from time, so we cannot think of *any* object apart from the possibility of its connexion with other things.

     If I can think of an object in the context of an atomic fact, I cannot think of it apart from the *possibility* of this context.

2.0122   The thing is independent, in so far as it can occur in all *possible* circumstances, but this form of independence is a form of connexion with the atomic fact, a form of dependence. (It is impossible for words to occur in two different ways, alone and in the proposition.)

2.0123   If I know an object, then I also know all the possibilities of its occurrence in atomic facts.

     (Every such possibility must lie in the nature of the object.)

     A new possibility cannot subsequently be found.

2.01231   In order to know an object, I must know not its external but all its internal qualities.

2.0124   If all objects are given, then thereby are all *possible* atomic facts also given.

2.013   Everything is, as it were, in a space of possible atomic facts. I can think of this space as empty, but not of the thing without the space.

2.0131   A spatial object must lie in infinite space. (A point in space is a place for an argument.)

A speck in a visual field need not be red, but it must have a colour; it has, so to speak, a colour space round it. A tone must have *a* pitch, the object of the sense of touch *a* hardness, etc.

2.014   Objects contain the possibility of all states of affairs.

2.0141   The possibility of its occurrence in atomic facts is the form of the object.

2.02   The object is simple.

2.0201   Every statement about complexes can be analysed into a statement about their constituent parts, and into those propositions which completely describe the complexes.

2.021   Objects form the substance of the world. Therefore they cannot be compound.

2.0211   If the world had no substance, then whether a proposition had sense would depend on whether another proposition was true.

2.0212   It would then be impossible to form a picture of the world (true or false).

2.022   It is clear that however different from the real one an imagined world may be, it must have something—a form—in common with the real world.

2.023   This fixed form consists of the objects.

2.0231   The substance of the world *can* only determine a form and not any material properties. For these are first presented by the propositions—first formed by the configuration of the objects.

2.0232   Roughly speaking: objects are colourless.

2.0233   Two objects of the same logical form are—apart from their external properties—only differentiated from one another in that they are different.

2.02331   Either a thing has properties which no other has, and then one can distinguish it straight away from the others by a description and refer to it; or, on the other hand, there are several things which have the totality of their properties in common, and then it is quite impossible to point to any one of them.

For if a thing is not distinguished by anything, I cannot distinguish it—for otherwise it would be distinguished.

2.024   Substance is what exists independently of what is the case.

2.025   It is form and content.

2.0251   Space, time and colour (colouredness) are forms of objects.

2.026   Only if there are objects can there be a fixed form of the world.

2.027   The fixed, the existent and the object are one.

2.0271   The object is the fixed, the existent; the configuration is the changing, the variable.

2.0272   The configuration of the objects forms the atomic fact.

2.03   In the atomic fact objects hang one in another, like the members of a chain.

2.031   In the atomic fact the objects are combined in a definite way.

2.032   The way in which objects hang together in the atomic fact is the structure of the atomic fact.

2.033   The form is the possibility of the structure.

2.034   The structure of the fact consists of the structures of the atomic facts.

2.04   The totality of existent atomic facts is the world.

2.05   The totality of existent atomic facts also determines which atomic facts do not exist.

2.06 The existence and nonexistence of atomic facts is the reality.

   (The existence of atomic facts we also call a positive fact, their nonexistence a negative fact.)

2.061 Atomic facts are independent of one another.

2.062 From the existence or nonexistence of an atomic fact we cannot infer the existence or nonexistence of another.

2.063 The total reality is the world.

2.1 We make to ourselves pictures of facts.

2.11 The picture presents the facts in logical space, the existence and nonexistence of atomic facts.

2.12 The picture is a model of reality.

2.13 To the objects correspond in the picture the elements of the picture.

2.131 The elements of the picture stand, in the picture, for the objects.

2.14 The picture consists in the fact that its elements are combined with one another in a definite way.

2.141 The picture is a fact.

2.15 That the elements of the picture are combined with one another in a definite way, represents that the things are so combined with one another.

   This connexion of the elements of the picture is called its structure, and the possibility of this structure is called the form of representation of the picture.

2.151 The form of representation is the possibility that the things are combined with one another as are the elements of the picture.

2.1511 Thus the picture is linked with reality; it reaches up to it.

2.1512 It is like a scale applied to reality.

2.15121 Only the outermost points of the dividing lines *touch* the object to be measured.

2.1513   According to this view the representing relation which makes it a picture, also belongs to the picture.

2.1514   The representing relation consists of the coordinations of the elements of the picture and the things.

2.1515   These coordinations are as it were the feelers of its elements with which the picture touches reality.

2.16   In order to be a picture a fact must have something in common with what it pictures.

2.161   In the picture and the pictured there must be something identical in order that the one can be a picture of the other at all.

2.17   What the picture must have in common with reality in order to be able to represent it after its manner—rightly or falsely—is its form of representation.

2.171   The picture can represent every reality whose form it has.

         The spatial picture, everything spatial, the coloured, everything coloured, etc.

2.172   The picture, however, cannot represent its form of representation; it shows it forth.

2.173   The picture represents its object from without (its standpoint is its form of representation), therefore the picture represents its object rightly or falsely.

2.174   But the picture cannot place itself outside of its form of representation.

2.18   What every picture, of whatever form, must have in common with reality in order to be able to represent it at all—rightly or falsely—is the logical form, that is, the form of reality.

2.181   If the form of representation is the logical form, then the picture is called a logical picture.

2.182   Every picture is *also* a logical picture. (On the other hand, for example, not every picture is spatial.)

2.19   The logical picture can depict the world.

2.2	The picture has the logical form of representation in common with what it pictures.
2.201	The picture depicts reality by representing a possibility of the existence and nonexistence of atomic facts.
2.202	The picture represents a possible state of affairs in logical space.
2.203	The picture contains the possibility of the state of affairs which it represents.
2.21	The picture agrees with reality or not; it is right or wrong, true or false.
2.22	The picture represents what it represents, independently of its truth or falsehood, through the form of representation.
2.221	What the picture represents is its sense.
2.222	In the agreement or disagreement of its sense with reality, its truth or falsity consists.
2.223	In order to discover whether the picture is true or false we must compare it with reality.
2.224	It cannot be discovered from the picture alone whether it is true or false.
2.225	There is no picture which is a priori true.
3	The logical picture of the facts is the thought.
3.001	"An atomic fact is thinkable"—means: we can imagine it.
3.01	The totality of true thoughts is a picture of the world.
3.02	The thought contains the possibility of the state of affairs which it thinks. What is thinkable is also possible.
3.03	We cannot think anything unlogical, for otherwise we should have to think unlogically.
3.031	It used to be said that God could create everything, except what was contrary to the laws of logic. The truth is, we could not *say* of an "unlogical" world how it would look.

3.032   To present in language anything which "contradicts logic" is as impossible as in geometry to present by its coordinates a figure which contradicts the laws of space; or to give the coordinates of a point which does not exist.

3.0321   We could present spatially an atomic fact which contradicted the laws of physics, but not one which contradicted the laws of geometry.

3.04   An a priori true thought would be one whose possibility guaranteed its truth.

3.05   We could only know a priori that a thought is true if its truth was to be recognized from the thought itself (without an object of comparison).

3.1   In the proposition the thought is expressed perceptibly through the senses.

3.11   We use the sensibly perceptible sign (sound or written sign, etc.) of the proposition as a projection of the possible state of affairs.

          The method of projection is the thinking of the sense of the proposition.

3.12   The sign through which we express the thought I call the propositional sign. And the proposition is the propositional sign in its projective relation to the world.

3.13   To the proposition belongs everything which belongs to the projection; but not what is projected.

          Therefore the possibility of what is projected but not this itself.

          In the proposition, therefore, its sense is not yet contained, but the possibility of expressing it.

          ("The content of the proposition" means the content of the significant proposition.)

          In the proposition the form of its sense is contained, but not its content.

3.14   The propositional sign consists in the fact that its elements, the words, are combined in it in a definite way.

The propositional sign is a fact.

3.141   The proposition is not a mixture of words (just as the musical theme is not a mixture of tones).

The proposition is articulate.

3.142   Only facts can express a sense, a class of names cannot.

3.143   That the propositional sign is a fact is concealed by the ordinary form of expression, written or printed.

(For in the printed proposition, for example, the sign of a proposition does not appear essentially different from a word. Thus it was possible for Frege to call the proposition a compounded name.)

3.1431   The essential nature of the propositional sign becomes very clear when we imagine it made up of spatial objects (such as tables, chairs, books) instead of written signs.

The mutual spatial position of these things then expresses the sense of the proposition.

3.1432   We must not say, "The complex sign 'aRb' says 'a stands in relation R to b'"; but we must say, "That 'a' stands in a certain relation to 'b' says that aRb."

3.144   States of affairs can be described but not *named*.

(Names resemble points; propositions resemble arrows, they have sense.)

3.2   In propositions thoughts can be so expressed that to the objects of the thoughts correspond the elements of the propositional sign.

3.201   These elements I call "simple signs" and the proposition "completely analysed."

3.202   The simple signs employed in propositions are called names.

3.203   The name means the object. The object is its meaning. ("A" is the same sign as "A.")

3.21   To the configuration of the simple signs in the proposi-
         tional sign corresponds the configuration of the objects
         in the state of affairs.

3.22   In the proposition the name represents the object.

3.221   Objects I can only *name*. Signs represent them. I can
         only speak *of* them. I cannot *assert them*. A proposition
         can only say *how* a thing is, not *what* it is.

3.23   The postulate of the possibility of the simple signs is the
         postulate of the determinateness of the sense.

3.24   A proposition about a complex stands in internal rela-
         tion to the proposition about its constituent part.

         A complex can only be given by its description, and
         this will either be right or wrong. The proposition in
         which there is mention of a complex, if this does not
         exist, becomes not nonsense but simply false.

         That a propositional element signifies a complex can
         be seen from an indeterminateness in the propositions
         in which it occurs. We *know* that everything is not yet
         determined by this proposition. (The notation for gen-
         erality *contains* a prototype.)

         The combination of the symbols of a complex in a
         simple symbol can be expressed by a definition.

3.25   There is one and only one complete analysis of the
         proposition.

3.251   The proposition expresses what it expresses in a definite
         and clearly specifiable way: the proposition is articulate.

3.26   The name cannot be analysed further by any definition.
         It is a primitive sign.

3.261   Every defined sign signifies *via* those signs by which it is
         defined, and the definitions show the way.

         Two signs, one a primitive sign, and one defined by
         primitive signs, cannot signify in the same way. Names
         cannot be taken to pieces by definition (nor any sign
         which alone and independently has a meaning).

3.262   What does not get expressed in the sign is shown by its application. What the signs conceal, their application declares.

3.263   The meanings of primitive signs can be explained by elucidations. Elucidations are propositions which contain the primitive signs. They can, therefore, only be understood when the meanings of these signs are already known.

3.3   Only the proposition has sense; only in the context of a proposition has a name meaning.

3.31   Every part of a proposition which characterizes its sense I call an expression (a symbol).

        (The proposition itself is an expression.)

        Expressions are everything—essential for the sense of the proposition—that propositions can have in common with one another.

        An expression characterizes a form and a content.

3.311   An expression presupposes the forms of all propositions in which it can occur. It is the common characteristic mark of a class of propositions.

3.312   It is therefore represented by the general form of the propositions which it characterizes.

        And in this form the expression is *constant* and everything else *variable*.

3.313   An expression is thus presented by a variable, whose values are the propositions which contain the expression.

        (In the limiting case the variables become constants, the expression a proposition.)

        I call such a variable a "propositional variable."

3.314   An expression has meaning only in a proposition. Every variable can be conceived as a propositional variable.

        (Including the variable name.)

3.315   If we change a constituent part of a proposition into a variable, there is a class of propositions which are all the values of the resulting variable proposition. This class in general still depends on what, by arbitrary agreement, we mean by parts of that proposition. But if we change all those signs, whose meaning was arbitrarily determined, into variables, there always remains such a class. But this is now no longer dependent on any agreement; it depends only on the nature of the proposition. It corresponds to a logical form, to a logical prototype.

3.316   What values the propositional variable can assume is determined.

The determination of the values *is* the variable.

3.317   The determination of the values of the propositional variable is done by *indicating the propositions* whose common mark the variable is.

The determination is a description of these propositions.

The determination will therefore deal only with symbols not with their meaning.

And *only* this is essential to the determination, *that it is only a description of symbols and asserts nothing about what is symbolized.*

The way in which we describe the propositions is not essential.

3.318   I conceive the proposition—like Frege and Russell—as a function of the expressions contained in it.

3.32   The sign is the part of the symbol perceptible by the senses.

3.321   Two different symbols can therefore have the sign (the written sign or the sound sign) in common—they then signify in different ways.

3.322   It can never indicate the common characteristic of two objects that we symbolize them with the same signs but by different *methods of symbolizing*. For the sign is arbitrary. We could therefore equally well choose two different signs and where then would be what was common in the symbolization.

3.323   In the language of everyday life it very often happens that the same word signifies in two different ways—and therefore belongs to two different symbols—or that two words, which signify in different ways, are apparently applied in the same way in the proposition.

Thus the word "is" appears as the copula, as the sign of equality, and as the expression of existence; "to exist" as an intransitive verb like "to go"; "identical" as an adjective; we speak of *something* but also of the fact of *something* happening.

(In the proposition "Green is green"—where the first word is a proper name and the last an adjective—these words have not merely different meanings but they are *different symbols*.)

3.324   Thus there easily arise the most fundamental confusions (of which the whole of philosophy is full).

3.325   In order to avoid these errors, we must employ a symbolism which excludes them, by not applying the same sign in different symbols and by not applying signs in the same way which signify in different ways. A symbolism, that is to say, which obeys the rules of *logical* grammar—of logical syntax.

(The logical symbolism of Frege and Russell is such a language, which, however, does still not exclude all errors.)

3.326   In order to recognize the symbol in the sign we must consider the significant use.

3.327   The sign determines a logical form only together with its logical syntactic application.

3.328   If a sign is *not necessary* then it is meaningless. That is the meaning of Occam's razor.

(If everything in the symbolism works as though a sign had meaning, then it has meaning.)

3.33   In logical syntax the meaning of a sign ought never to play a rôle; it must admit of being established without mention being thereby made of the *meaning* of a sign; it ought to presuppose *only* the description of the expressions.

3.331   From this observation we get a further view—into Russell's *Theory of Types*. Russell's error is shown by the fact that in drawing up his symbolic rules he has to speak of the meaning of the signs.

3.332   No proposition can say anything about itself, because the propositional sign cannot be contained in itself (that is the "whole theory of types").

3.333   A function cannot be its own argument, because the functional sign already contains the prototype of its own argument and it cannot contain itself.

     If, for example, we suppose that the function $F(fx)$ could be its own argument, then there would be a proposition "$F(F(fx))$," and in this the outer function F and the inner function F must have different meanings; for the inner has the form $\phi(fx)$, the outer the form $\psi(\phi(fx))$. Common to both functions is only the letter "F," which by itself signifies nothing.

     This is at once clear, if instead of "$F(F(u))$" we write "$(\exists\phi) : F(\phi u). \phi u = Fu$."

     Herewith Russell's paradox vanishes.

3.334   The rules of logical syntax must follow of themselves, if we only know how every single sign signifies.

3.34   A proposition possesses essential and accidental features.

     Accidental are the features which are due to a particular way of producing the propositional sign. Essential are those which alone enable the proposition to express its sense.

3.341   The essential in a proposition is therefore that which is common to all propositions which can express the same sense.

      And in the same way in general the essential in a symbol is that which all symbols which can fulfil the same purpose have in common.

3.3411   One could therefore say the real name is that which all symbols, which signify an object, have in common. It would then follow, step by step, that no sort of composition was essential for a name.

3.342   In our notations there is indeed something arbitrary, but *this* is not arbitrary, namely that *if* we have determined anything arbitrarily, then something else *must* be the case. (This results from the *essence* of the notation.)

3.3421   A particular method of symbolizing may be unimportant, but it is always important that this is a *possible* method of symbolizing. And this happens as a rule in philosophy: The single thing proves over and over again to be unimportant, but the possibility of every single thing reveals something about the nature of the world.

3.343   Definitions are rules for the translation of one language into another. Every correct symbolism must be translatable into every other according to such rules. It is *this* which all have in common.

3.344   What signifies in the symbol is what is common to all those symbols by which it can be replaced according to the rules of logical syntax.

3.3441   We can, for example, express what is common to all notations for the truth-functions as follows: It is common to them that they all, for example, *can be replaced* by the notations of "$\sim p$" ("not p") and "$p \vee q$" ("p or q").

      (Herewith is indicated the way in which a special possible notation can give us general information.)

3.3442   The sign of the complex is not arbitrarily resolved in the analysis, in such a way that its resolution would be different in every propositional structure.

3.4   The proposition determines a place in logical space: the existence of this logical place is guaranteed by the existence of the constituent parts alone, by the existence of the significant proposition.

3.41   The propositional sign and the logical coordinates: that is the logical place.

3.411   The geometrical and the logical place agree in that each is the possibility of an existence.

3.42   Although a proposition may only determine one place in logical space, the whole logical space must already be given by it.

     (Otherwise denial, the logical sum, the logical product, etc., would always introduce new elements—in co-ordination.)

     (The logical scaffolding round the picture determines the logical space. The proposition reaches through the whole logical space.)

3.5   The applied, thought, propositional sign is the thought.

4   The thought is the significant proposition.

4.001   The totality of propositions is the language.

4.002   Man possesses the capacity of constructing languages, in which every sense can be expressed, without having an idea how and what each word means—just as one speaks without knowing how the single sounds are produced.

     Colloquial language is a part of the human organism and is not less complicated than it.

     From it it is humanly impossible to gather immediately the logic of language.

     Language disguises the thought; so that from the external form of the clothes one cannot infer the form of the thought they clothe, because the external form of the clothes is constructed with quite another object than to let the form of the body be recognized.

The silent adjustments to understand colloquial language are enormously complicated.

4.003 Most propositions and questions, that have been written about philosophical matters, are not false, but senseless. We cannot, therefore, answer questions of this kind at all, but only state their senselessness. Most questions and propositions of the philosophers result from the fact that we do not understand the logic of our language.

(They are of the same kind as the question whether the Good is more or less identical than the Beautiful.)

And so it is not to be wondered at that the deepest problems are really *no* problems.

4.0031 All philosophy is "Critique of language" (but not at all in Mauthner's sense). Russell's merit is to have shown that the apparent logical form of the proposition need not be its real form.

4.01 The proposition is a picture of reality.

The proposition is a model of the reality as we think it is.

4.011 At the first glance the proposition—say as it stands printed on paper—does not seem to be a picture of the reality of which it treats. But nor does the musical score appear at first sight to be a picture of a musical piece; nor does our phonetic spelling (letters) seem to be a picture of our spoken language. And yet these symbolisms prove to be pictures—even in the ordinary sense of the word—of what they represent.

4.012 It is obvious that we perceive a proposition of the form *aRb* as a picture. Here the sign is obviously a likeness of the signified.

4.013   And if we penetrate to the essence of this pictorial nature we see that this is not disturbed by *apparent irregularities* (like the use of ♯ and ♭ in the score).

   For these irregularities also picture what they are to express; only in another way.

4.014   The gramophone record, the musical thought, the score, the waves of sound, all stand to one another in that pictorial internal relation, which holds between language and the world. To all of them the logical structure is common.

   (Like the two youths, their two horses and their lilies in the story. They are all in a certain sense one.)

4.0141   In the fact that there is a general rule by which the musician is able to read the symphony out of the score, and that there is a rule by which one could reconstruct the symphony from the line on a gramophone record and from this again—by means of the first rule—construct the score, herein lies the internal similarity between these things which at first sight seem to be entirely different. And the rule is the law of projection which projects the symphony into the language of the musical score. It is the rule of translation of this language into the language of the gramophone record.

4.015   The possibility of all similes, of all the imagery of our language, rests on the logic of representation.

4.016   In order to understand the essence of the proposition, consider hieroglyphic writing, which pictures the facts it describes.

   And from it came the alphabet without the essence of the representation being lost.

4.02   This we see from the fact that we understand the sense of the propositional sign, without having had it explained to us.

4.021   The proposition is a picture of reality, for I know the state of affairs presented by it, if I understand the proposition. And I understand the proposition, without its sense having been explained to me.

4.022   The proposition *shows* its sense.

  The proposition *shows* how things stand, *if* it is true. And it *says*, that they do so stand.

4.023   The proposition determines reality to this extent, that one only needs to say "Yes" or "No" to it to make it agree with reality.

  It must therefore be completely described by the proposition.

  A proposition is the description of a fact.

  As the description of an object describes it by its external properties so propositions describe reality by its internal properties.

  The proposition constructs a world with the help of a logical scaffolding, and therefore one can actually see in the proposition all the logical features possessed by reality if it is true. One can *draw conclusions* from a false proposition.

4.024   To understand a proposition means to know what is the case, if it is true.

  (One can therefore understand it without knowing whether it is true or not.)

  One understands it if one understands its constituent parts.

4.025   The translation of one language into another is not a process of translating each proposition of the one into a proposition of the other, but only the constituent parts of propositions are translated.

  (And the dictionary does not only translate substantives but also adverbs and conjunctions, etc., and it treats them all alike.)

4.026   The meanings of the simple signs (the words) must be explained to us, if we are to understand them.

By means of propositions we explain ourselves.

4.027   It is essential to propositions, that they can communicate a *new* sense to us.

4.03   A proposition must communicate a new sense with old words.

The proposition communicates to us a state of affairs, therefore it must be *essentially* connected with the state of affairs.

And the connexion is, in fact, that it is its logical picture.

The proposition only asserts something, in so far as it is a picture.

4.031   In the proposition a state of affairs is, as it were, put together for the sake of experiment.

One can say, instead of, This proposition has such and such a sense, This proposition represents such and such a state of affairs.

4.0311 One name stands for one thing, and another for another thing, and they are connected together. And so the whole, like a living picture, presents the atomic fact.

4.0312 The possibility of propositions is based upon the principle of the representation of objects by signs.

My fundamental thought is that the "logical constants" do not represent. That the *logic* of the facts cannot be represented.

4.032   The proposition is a picture of its state of affairs, only in so far as it is logically articulated.

(Even the proposition "ambulo" is composite, for its stem gives a different sense with another termination, or its termination with another stem.)

4.04   In the proposition there must be exactly as many things distinguishable as there are in the state of affairs, which it represents.

They must both possess the same logical (mathematical) multiplicity (cf. Hertz's *Mechanics*, on Dynamic Models).

4.041 This mathematical multiplicity naturally cannot in its turn be represented. One cannot get outside it in the representation.

4.0411 If we tried, for example, to express what is expressed by "$(x).fx$" by putting an index before fx, like: "Gen. fx," it would not do, we should not know what was generalized. If we tried to show it by an index g, like: "$f(x_g)$" it would not do—we should not know the scope of the generalization.

If we were to try it by introducing a mark in the argument places, like "$(G,G).F\ (G,G)$," it would not do—we could not determine the identity of the variables, etc.

All these ways of symbolizing are inadequate because they have not the necessary mathematical multiplicity.

4.0412 For the same reason the idealist explanation of the seeing of spatial relations through "spatial spectacles" does not do, because it cannot explain the multiplicity of these relations.

4.05 Reality is compared with the proposition.

4.06 Propositions can be true or false only by being pictures of the reality.

4.061 If one does not observe that propositions have a sense independent of the facts, one can easily believe that true and false are two relations between signs and things signified with equal rights.

One could then, for example, say that "p" signifies in the true way what "$\sim p$" signifies in the false way, etc.

4.062   Can we not make ourselves understood by means of false propositions as hitherto with true ones, so long as we know that they are meant to be false? No! For a proposition is true, if what we assert by means of it is the case; and if by "*p*" we mean ~*p*, and what we mean is the case, then "*p*" in the new conception is true and not false.

4.0621   That, however, the signs "*p*" and "~*p*" *can* say the same thing is important, for it shows that the sign "~" corresponds to nothing in reality.

That negation occurs in a proposition, is no characteristic of its sense ($\sim \sim p = p$).

The propositions "*p*" and "~*p*" have opposite senses, but to them corresponds one and the same reality.

4.063   An illustration to explain the concept of truth. A black spot on white paper; the form of the spot can be described by saying of each point of the plane whether it is white or black. To the fact that a point is black corresponds a positive fact; to the fact that a point is white (not black), a negative fact. If I indicate a point of the plane (a truth-value in Frege's terminology), this corresponds to the assumption proposed for judgment, etc. etc.

But to be able to say that a point is black or white, I must first know under what conditions a point is called white or black; in order to be able to say "*p*" is true (or false) I must have determined under what conditions I call "*p*" true, and thereby I determine the sense of the proposition.

The point at which the simile breaks down is this: we can indicate a point on the paper, without knowing what white and black are; but to a proposition without a sense corresponds nothing at all, for it signifies no thing (truth-value) whose properties are called "false" or "true"; the verb of the proposition is not "is true" or "is false"—as Frege thought—but that which "is true" must already contain the verb.

4.064   Every proposition must *already* have a sense; assertion cannot give it a sense, for what it asserts is the sense itself. And the same holds of denial, etc.

4.0641 One could say, the denial is already related to the logical place determined by the proposition that is denied.

The denying proposition determines a logical place *other* than does the proposition denied.

The denying proposition determines a logical place, with the help of the logical place of the proposition denied, by saying that it lies outside the latter place.

That one can deny again the denied proposition, shows that what is denied is already a proposition and not merely the preliminary to a proposition.

4.1   A proposition presents the existence and nonexistence of atomic facts.

4.11   The totality of true propositions is the total natural science (or the totality of the natural sciences).

4.111   Philosophy is not one of the natural sciences.

(The word "*philosophy*" must mean something which stands above or below, but not beside the natural sciences.)

4.112   The object of philosophy is the logical clarification of thoughts.

Philosophy is not a theory but an activity.

A philosophical work consists essentially of elucidations.

The result of philosophy is not a number of "philosophical propositions," but to make propositions clear.

Philosophy should make clear and delimit sharply the thoughts which otherwise are, as it were, opaque and blurred.

4.1121   Psychology is no nearer related to philosophy, than is any other natural science.

The theory of knowledge is the philosophy of psychology.

Does not my study of sign language correspond to the study of thought processes which philosophers held to be so essential to the philosophy of logic? Only they got entangled for the most part in unessential psychological investigations, and there is an analogous danger for my method.

4.1122   The Darwinian theory has no more to do with philosophy than has any other hypothesis of natural science.

4.113   Philosophy limits the disputable sphere of natural science.

4.114   It should limit the thinkable and thereby the unthinkable.

It should limit the unthinkable from within through the thinkable.

4.115   It will mean the unspeakable by clearly displaying the speakable.

4.116   Everything that can be thought at all can be thought clearly. Everything that can be said can be said clearly.

4.12   Propositions can represent the whole reality, but they cannot represent what they must have in common with reality in order to be able to represent it—the logical form.

To be able to represent the logical form, we should have to be able to put ourselves with the propositions outside logic, that is outside the world.

4.121   Propositions cannot represent the logical form: this mirrors itself in the propositions.

That which mirrors itself in language, language cannot represent.

That which expresses *itself* in language, *we* cannot express by language.

The propositions *show* the logical form of reality.

They exhibit it.

4.1211   Thus a proposition "*fa*" shows that in its sense the object *a* occurs, two propositions "*fa*" and "*ga*" that they are both about the same object.

If two propositions contradict one another, this is shown by their structure; similarly if one follows from another, etc.

4.1212   What *can* be shown *cannot* be said.

4.1213   Now we understand our feeling that we are in possession of the right logical conception, if only all is right in our symbolism.

4.122   We can speak in a certain sense of formal properties of objects and atomic facts, or of properties of the structure of facts, and in the same sense of formal relations and relations of structures.

(Instead of property of the structure I also say "internal property"; instead of relation of structures "internal relation.")

I introduce these expressions in order to show the reason for the confusion, very widespread among philosophers, between internal relations and proper [external] relations.)

The holding of such internal properties and relations cannot, however, be asserted by propositions, but it shows itself in the propositions, which present the atomic facts and treat of the objects in question.

4.1221   An internal property of a fact we also call a feature of this fact. (In the sense in which we speak of facial features.)

4.123   A property is internal if it is unthinkable that its object does not possess it.

(This blue colour and that stand in the internal relation of brighter and darker eo ipso. It is unthinkable that *these* two objects should not stand in this relation.)

(Here to the shifting use of the words "property" and "relation" there corresponds the shifting use of the word "object.")

4.124   The existence of an internal property of a possible state of affairs is not expressed by a proposition, but it expresses itself in the proposition which presents that state of affairs, by an internal property of this proposition.

It would be as senseless to ascribe a formal property to a proposition as to deny it the formal property.

4.1241   One cannot distinguish forms from one another by saying that one has this property but the other that: for this assumes that there is a sense in asserting either property of either form.

4.125   The existence of an internal relation between possible states of affairs expresses itself in language by an internal relation between the propositions presenting them.

4.1251   Here the disputed question "whether all relations are internal or external" disappears.

4.1252   Series which are ordered by *internal* relations I call formal series.

The series of numbers is ordered not by an external, but by an internal relation.

Similarly the series of propositions "*aRb*,"

"$(\exists x) : aRx . xRb$,"

"$(\exists x,y) : aRx . xRy . yRb$," etc.

(If *b* stands in one of these relations to *a*, I call *b* a successor of *a*.)

4.126   In the sense in which we speak of formal properties we can now speak also of formal concepts.

(I introduce this expression in order to make clear the confusion of formal concepts with proper concepts which runs through the whole of the old logic.)

That anything falls under a formal concept as an object belonging to it, cannot be expressed by a proposition. But it shows itself in the sign of this object itself. (The name shows that it signifies an object, the numerical sign that it signifies a number, etc.)

Formal concepts cannot, like proper concepts, be presented by a function.

For their characteristics, the formal properties, are not expressed by the functions.

The expression of a formal property is a feature of certain symbols.

The sign that signifies the characteristics of a formal concept is, therefore, a characteristic feature of all symbols, whose meanings fall under the concept.

The expression of the formal concept is therefore a propositional variable in which only this characteristic feature is constant.

4.127   The propositional variable signifies the formal concept, and its values signify the objects which fall under this concept.

4.1271   Every variable is the sign of a formal concept.

For every variable presents a constant form, which all its values possess, and which can be conceived as a formal property of these values.

4.1272   So the variable name "x" is the proper sign of the pseudo-concept *object*.

Wherever the word "object" ("thing," "entity," etc.) is rightly used, it is expressed in logical symbolism by the variable name.

For example in the proposition "there are two objects which . . . ," by "$(\exists x,y) \ldots$."

Wherever it is used otherwise, i.e., as a proper concept word, there arise senseless pseudo-propositions.

So one cannot, e.g., say "There are objects" as one says "There are books." Nor "There are 100 objects" or "There are ℵ₀ objects." And it is senseless to speak of the *number of all objects*.

The same holds of the words "Complex," "Fact," "Function," "Number," etc.

They all signify formal concepts and are presented in logical symbolism by variables, not by functions or classes (as Frege and Russell thought).

Expressions like "1 is a number," "there is only one number nought," and all like them are senseless.

(It is as senseless to say, "there is only one 1" as it would be to say: 2 + 2 is at 3 o'clock equal to 4.)

4.12721 The formal concept is already given with an object, which falls under it. One cannot, therefore, introduce both, the objects which fall under a formal concept *and* the formal concept itself, as primitive ideas. One cannot, therefore, e.g., introduce (as Russell does) the concept of function and also special functions as primitive ideas; or the concept of number and definite numbers.

4.1273 If we want to express in logical symbolism the general proposition "*b* is a successor of *a*" we need for this an expression for the general term of the formal series: aRb, $(\exists x): aRx.\ xRb$, $(\exists x,y): aRx.\ xRy.\ yRb, \ldots$ The general term of a formal series can only be expressed by a variable, for the concept symbolized by "term of this formal series" is a *formal* concept. (This Frege and Russell overlooked; the way in which they express general propositions like the above is, therefore, false; it contains a vicious circle.)

We can determine the general term of the formal series by giving its first term and the general form of the operation, which generates the following term out of the preceding proposition.

4.1274 The question about the existence of a formal concept is senseless. For no proposition can answer such a question.

(For example, one cannot ask: "Are there unanalysable subject-predicate propositions?")

4.128 The logical forms are *anumerical*.

Therefore there are in logic no preeminent numbers, and therefore there is no philosophical monism or dualism, etc.

4.2 The sense of a proposition is its agreement and disagreement with the possibilities of the existence and nonexistence of the atomic facts.

4.21 The simplest proposition, the elementary proposition, asserts the existence of an atomic fact.

4.211 It is a sign of an elementary proposition, that no elementary proposition can contradict it.

4.22 The elementary proposition consists of names. It is a connexion, a concatenation, of names.

4.221 It is obvious that in the analysis of propositions we must come to elementary propositions, which consist of names in immediate combination.

The question arises here, how the propositional connexion comes to be.

4.2211 Even if the world is infinitely complex, so that every fact consists of an infinite number of atomatic facts and every atomic fact is composed of an infinite number of objects, even then there must be objects and atomic facts.

4.23 The name occurs in the proposition only in the context of the elementary proposition.

4.24 The names are the simple symbols, I indicate them by single letters (x, y, z).

The elementary proposition I write as function of the names, in the form "fx," "$\phi(x,y)$," etc.

Or I indicate it by the letters p,q,r.

4.241   If I use two signs with one and the same meaning, I express this by putting between them the sign "$=$."

"$a = b$" means then, that the sign "a" is replaceable by the sign "b."

(If I introduce by an equation a new sign "b," by determining that it shall replace a previously known sign "a," I write the equation—definition—(like Russell) in the form "$a = b$ Def." A definition is a symbolic rule.)

4.242   Expressions of the form "$a = b$" are therefore only expedients in presentation: They assert nothing about the meaning of the signs "a" and "b."

4.243   Can we understand two names without knowing whether they signify the same thing or two different things? Can we understand a proposition in which two names occur, without knowing if they mean the same or different things?

If I know the meaning of an English and a synonymous German word, it is impossible for me not to know that they are synonymous, it is impossible for me not to be able to translate them into one another.

Expressions like "$a = a$," or expressions deduced from these are neither elementary propositions nor otherwise significant signs. (This will be shown later.)

4.25   If the elementary proposition is true, the atomic fact exists; if it is false the atomic fact does not exist.

4.26   The specification of all true elementary propositions describes the world completely. The world is completely described by the specification of all elementary propositions plus the specification, which of them are true and which false.

4.27   With regard to existence of n atomic facts there are

$$K_n = \sum_{v=0}^{n} \binom{n}{v} \text{ possibilities.}$$

It is possible for all combinations of atomic facts to exist, and the others not to exist.

4.28   To these combinations correspond the same number of possibilities of the truth—and falsehood—of n elementary propositions.

4.3   The truth-possibilities of the elementary propositions mean the possibilities of the existence and nonexistence of the atomic facts.

4.31   The truth-possibilities can be presented by schemata of the following kind ("T" means "true," "F" "false." The rows of T's and F's under the row of the elementary propositions mean their truth-possibilities in an easily intelligible symbolism).

p	q	r
T	T	T
F	T	T
T	F	T
T	T	F
F	F	T
F	T	F
T	F	F
F	F	F

p	q
T	T
F	T
T	F
F	F

p
T
F

4.4   A proposition is the expression of agreement and disagreement with the truth-possibilities of the elementary propositions.

4.41   The truth-possibilities of the elementary propositions
        are the conditions of the truth and falsehood of the
        propositions.

4.411   It seems probable even at first sight that the introduction
        of the elementary propositions is fundamental for the
        comprehension of the other kinds of propositions. In-
        deed the comprehension of the general propositions de-
        pends *palpably* on that of the elementary propositions.

4.42   With regard to the agreement and disagreement of a
        proposition with the truth-possibilities of n elementary

        propositions there are $\sum_{\kappa=0}^{K_n} \binom{K_n}{\kappa} = L_n$ possibilities.

4.43   Agreement with the truth-possibilities can be expressed
        by coordinating with them in the schema the mark "T"
        (true).

        Absence of this mark means disagreement.

4.431   The expression of the agreement and disagreement with
        the truth-possibilities of the elementary propositions
        expresses the truth-conditions of the proposition.

        The proposition is the expression of its truth-
        conditions.

        (Frege has therefore quite rightly put them at the
        beginning, as explaining the signs of his logical sym-
        bolism. Only Frege's explanation of the truth-concept
        is false: if "the true" and "the false" were real objects
        and the arguments in $\sim p$, etc., then the sense of $\sim p$
        would by no means be determined by Frege's determi-
        nation.)

4.44   The sign which arises from the coordination of that mark
        "T" with the truth-possibilities is a propositional sign.

4.441   It is clear that to the complex of the signs "F" and "T"
        no object (or complex of objects) corresponds; anymore

than to horizontal and vertical lines or to brackets. There are no "logical objects."

Something analogous holds of course for all signs, which express the same as the schemata of "T" *and* "F."

4.442　Thus e.g.,

"

p	q	
T	T	T
F	T	T
T	F	
F	F	T

"

is a propositional sign.

(Frege's assertion sign "⊢" is logically altogether meaningless; in Frege [and Russell] it only shows that these authors hold as true the propositions marked in this way.

"⊢" belongs therefore to the propositions no more than does the number of the proposition. A proposition cannot possibly assert of itself that it is true.)

If the sequence of the truth-possibilities in the schema is once for all determined by a rule of combination, then the last column is by itself an expression of the truth-conditions. If we write this column as a row the propositional sign becomes:

"(TT—T) (p,q)," or more plainly: "(TTFT) (p,q)."

(The number of places in the left-hand bracket is determined by the number of terms in the right-hand bracket.)

4.45　For n elementary propositions there are L_n possible groups of truth-conditions.

The groups of truth-conditions which belong to the truth-possibilities of a number of elementary propositions can be ordered in a series.

4.46   Among the possible groups of truth-conditions there
         are two extreme cases.

         In the one case the proposition is true for all the
         truth-possibilities of the elementary propositions. We say
         that the truth-conditions are *tautological.*

         In the second case the proposition is false for all
         the truth-possibilities. The truth-conditions are *self-
         contradictory.*

         In the first case we call the proposition a tautology,
         in the second case a contradiction.

4.461   The proposition shows what it says, the tautology and
         the contradiction that they say nothing.

         The tautology has no truth-conditions, for it is un-
         conditionally true; and the contradiction is on no con-
         dition true.

         Tautology and contradiction are without sense.

         (Like the point from which two arrows go out in op-
         posite directions.)

         (I know, e.g., nothing about the weather, when I
         know that it rains or does not rain.)

4.4611   Tautology and contradiction are, however, not sense-
         less; they are part of the symbolism, in the same way
         that "0" is part of the symbolism of Arithmetic.

4.462   Tautology and contradiction are not pictures of the
         reality. They present no possible state of affairs. For
         the one allows *every* possible state of affairs, the other
         none.

         In the tautology the conditions of agreement with the
         world—the presenting relations—cancel one another, so
         that it stands in no presenting relation to reality.

4.463   The truth-conditions determine the range, which is left
         to the facts by the proposition.

         (The proposition, the picture, the model, are in a
         negative sense like a solid body, which restricts the free

movement of another: in a positive sense, like the space limited by solid substance, in which a body may be placed.)

Tautology leaves to reality the whole infinite logical space; contradiction fills the whole logical space and leaves no point to reality. Neither of them, therefore, can in any way determine reality.

4.464   The truth of tautology is certain, of propositions possible, of contradiction impossible.

(Certain, possible, impossible: here we have an indication of that gradation which we need in the theory of probability.)

4.465   The logical product of a tautology and a proposition says the same as the proposition. Therefore that product is identical with the proposition. For the essence of the symbol cannot be altered without altering its sense.

4.466   To a definite logical combination of signs corresponds a definite logical combination of their meanings; *every arbitrary* combination only corresponds to the unconnected signs.

That is, propositions which are true for every state of affairs cannot be combinations of signs at all, for otherwise there could only correspond to them definite combinations of objects.

(And to no logical combination corresponds *no* combination of the objects.)

Tautology and contradiction are the limiting cases of the combinations of symbols, namely their dissolution.

4.4661   Of course the signs are also combined with one another in the tautatus and contradiction, i.e., they stand in relations to one another, but these relations are meaningless, unessential to the *symbol*.

4.5   Now it appears to be possible to give the most general form of proposition; i.e., to give a description of the propositions of some one sign language, so that every possible sense can be expressed by a symbol, which falls under the description, and so that every symbol which falls under the description can express a sense, if the meanings of the names are chosen accordingly.

It is clear that in the description of the most general form of proposition *only* what is essential to it may be described—otherwise it would not be the most general form.

That there is a general form is proved by the fact that there cannot be a proposition whose form could not have been foreseen (i.e., constructed). The general form of proposition is: Such and such is the case.

4.51   Suppose *all* elementary propositions were given me: then we can simply ask: what propositions I can build out of them. And these are *all* propositions and *so* are they limited.

4.52   The propositions are everything which follows from the totality of all elementary propositions (of course also from the fact that it is the *totality of them all*). (So, in some sense, one could say, that *all* propositions are generalizations of the elementary propositions.)

4.53   The general propositional form is a variable.

5   Propositions are truth-functions of elementary propositions.

(An elementary proposition is a truth-function of itself.)

5.01   The elementary propositions are the truth-arguments of propositions.

5.02   It is natural to confuse the arguments of functions with the indices of names. For I recognize the meaning of the sign containing it from the argument just as much as from the index.

In Russell's "$+_c$," for example, "c" is an index which

indicates that the whole sign is the addition sign for cardinal numbers. But this way of symbolizing depends on arbitrary agreement, and one could choose a simple sign instead of "$+_c$": but in "$\sim p$" "p" is not an index but an argument; the sense of "$\sim p$" *cannot* be understood, unless the sense of "p" has previously been understood. (In the name Julius Cæsar, Julius is an index. The index is always part of a description of the object to whose name we attach it, e.g., *The* Cæsar of the Julian gens.)

The confusion of argument and index is, if I am not mistaken, at the root of Frege's theory of the meaning of propositions and functions. For Frege the propositions of logic were names and their arguments the indices of these names.

5.1   The truth-functions can be ordered in series.

That is the foundation of the theory of probability.

5.101   The truth-functions of every number of elementary propositions can be written in a schema of the following kind:

(T T T T) (p, q) Tautology (if p then p, and if q then q)
$[p \supset p . q \supset q]$

(F T T T) (p, q) in words: Not both p and q. $[\sim(p . q)]$
(T F T T) (p, q) " " If q then p. $[q \supset p]$
(T T F T) (p, q) " " If p then q. $[p \supset q]$
(T T T F) (p, q) " " p or q. $[p \vee q]$
(F F T T) (p, q) " " Not q. $[\sim q]$
(F T F T) (p, q) " " Not p. $[\sim p]$
(F T T F) (p, q) " " p or q, but not both. $[p . \sim q : v : q . \sim p]$

(T F F T) (p, q) " " If p, then q; and if q, then p. $[p \equiv q]$
(T F T F) (p, q) " " p
(T T F F) (p, q) " " q
(F F F T) (p, q) " " Neither p nor q. $[\sim p . \sim q$ or $p | q]$

(F F T F) (p, q) in words: p and not q. $[p . \sim q]$

(F T F F) (p,q) "   " q and not p. $[q . \sim p]$

(T F F F) (p,q) "   " q and p. $[q.p]$

(F F F F) (p,q) Contradiction (p and not p; and q and not q.)

$$[p . \sim p . q . \sim q]$$

Those truth-possibilities of its truth-arguments, which verify the proposition, I shall call its *truth-grounds*.

5.11   If the truth-grounds which are common to a number of propositions are also truth-grounds of some one proposition, we say that the truth of this proposition follows from the truth of those propositions.

5.12   In particular the truth of a proposition p follows from that of a proposition q, if all the truth-grounds of the second are truth-grounds of the first.

5.121   The truth-grounds of q are contained in those of p; p follows from q.

5.122   If p follows from q, the sense of "p" is contained in that of "q."

5.123   If a god creates a world in which certain propositions are true, he creates thereby also a world in which all propositions consequent on them are true. And similarly he could not create a world in which the proposition "p" is true without creating all its objects.

5.124   A proposition asserts every proposition which follows from it.

5.1241   "$p.q$" is one of the propositions which assert "p" and at the same time one of the propositions which assert "q."

Two propositions are opposed to one another if there is no significant proposition which asserts them both.

Every proposition which contradicts another, denies it.

5.13   That the truth of one proposition follows from the truth of other propositions, we perceive from the structure of the propositions.

5.131   If the truth of one proposition follows from the truth of others, this expresses itself in relations in which the forms of these propositions stand to one another, and we do not need to put them in these relations first by connecting them with one another in a proposition; for these relations are internal, and exist as soon as, and by the very fact that, the propositions exist.

5.1311   When we conclude from p v q and $\sim p$ to q the relation between the forms of the propositions "p v q" and "$\sim p$" is here concealed by the method of symbolizing. But if we write, e.g., instead of "p v q" "$p|q$.|. $p|q$" and instead of "$\sim p$" "$p|p$" ($p|q$ = neither p nor q), then the inner connexion becomes obvious.

         (The fact that we can infer fa from (x). fx shows that generality is present also in the symbol "(x). fx."

5.132   If p follows from q, I can conclude from q to p; infer p from q.

         The method of inference is to be understood from the two propositions alone.

         Only they themselves can justify the inference.

         Laws of inference, which—as in Frege and Russell—are to justify the conclusions, are senseless and would be superfluous.

5.133   All inference takes place a priori.

5.134   From an elementary proposition no other can be inferred.

5.135   In no way can an inference be made from the existence of one state of affairs to the existence of another entirely different from it.

5.136   There is no causal nexus which justifies such an inference.

5.1361   The events of the future *cannot* be inferred from those of the present.

         Superstition is the belief in the causal nexus.

5.1362 The freedom of the will consists in the fact that future actions cannot be known now. We could only know them if causality were an *inner* necessity, like that of logical deduction.—The connexion of knowledge and what is known is that of logical necessity.

("A knows that *p* is the case" is senseless if *p* is a tautology.)

5.1363 If from the fact that a proposition is obvious to us it does not *follow* that it is true, then obviousness is no justification for our belief in its truth

5.14 If a proposition follows from another, then the latter says more than the former, the former less than the latter.

5.141 If *p* follows from *q* and *q* from *p* then they are one and the same proposition.

5.142 A tautology follows from all propositions: it says nothing.

5.143 Contradiction is something shared by propositions, which *no* proposition has in common with another. Tautology is that which is shared by all propositions, which have nothing in common with one another.

Contradiction vanishes so to speak outside, tautology inside all propositions.

Contradiction is the external limit of the propositions, tautology their substanceless centre.

5.15 If T_r is the number of the truth-grounds of the proposition "*r*," T_{rs} the number of those truth-grounds of the proposition "*s*" which are at the same time truth-grounds of "*r*," then we call the ratio $T_{rs} : T_r$ the measure of the *probability* which the proposition "*r*" gives to the proposition "*s*."

5.151 Suppose in a schema like that above in No. 5.101 T_r is the number of the "T"'s in the proposition *r*, T_{rs} the number of those "T"'s in the proposition *s*, which stand in the same columns as "*T*"'s of the proposition *r*; *then* the proposition r gives to the proposition s the probability $T_{rs} : T_r$.

5.1511 There is no special object peculiar to probability propositions.

5.152 Propositions which have no truth-arguments in common with one another we call independent.

Independent propositions (e.g., any two elementary propositions) give to one another the probability $\frac{1}{2}$.

If p follows from q, the proposition q gives to the proposition p the probability 1. The certainty of logical conclusion is a limiting case of probability.

(Application to tautology and contradiction.)

5.153 A proposition is in itself neither probable nor improbable. An event occurs or does not occur, there is no middle course.

5.154 In an urn there are equal numbers of white and black balls (and no others). I draw one ball after another and put them back in the urn. Then I can determine by the experiment that the numbers of the black and white balls which are drawn approximate as the drawing continues.

So *this* is not a mathematical fact.

If then, I say, It is equally probable that I should draw a white and a black ball, this means, All the circumstances known to me (including the natural laws hypothetically assumed) give to the occurrence of the one event no more probability than to the occurrence of the other. That is they give—as can easily be understood from the above explanations—to each the probability $\frac{1}{2}$.

What I can verify by the experiment is that the occurrence of the two events is independent of the circumstances, with which I have no closer acquaintance.

5.155   The unit of the probability proposition is: The circumstances—with which I am not further acquainted—give to the occurrence of a definite event such and such a degree of probability.

5.156   Probability is a generalization.

It involves a general description of a propositional form. Only in default of certainty do we need probability.

If we are not completely acquainted with a fact, but know *something* about its form.

(A proposition can, indeed, be an incomplete picture of a certain state of affairs, but it is always *a* complete picture.)

The probability proposition is, as it were, an extract from other propositions.

5.2   The structures of propositions stand to one another in internal relations.

5.21   We can bring out these internal relations in our manner of expression, by presenting a proposition as the result of an operation which produces it from other propositions (the bases of the operation).

5.22   The operation is the expression of a relation between the structures of its result and its bases.

5.23   The operation is that which must happen to a proposition in order to make another out of it.

5.231   And that will naturally depend on their formal properties, on the internal similarity of their forms.

5.232   The internal relation which orders a series is equivalent to the operation by which one term arises from another.

5.233   The first place in which an operation can occur is where a proposition arises from another in a logically significant way; i.e., where the logical construction of the proposition begins.

5.234   The truth-functions of elementary proposition are re-
        sults of operations which have the elementary proposi-
        tions as bases. (I call these operations, truth-operations.)

5.2341 The sense of a truth-function of p is a function of the
        sense of p.

        Denial, logical addition, logical multiplication, etc.
        etc., are operations.

        (Denial reverses the sense of a proposition.)

5.24   An operation shows itself in a variable; it shows how we
        can proceed from one form of proposition to another.

        It gives expression to the difference between the
        forms.

        (And that which is common to the bases, and the re-
        sult of an operation, is the bases themselves.)

5.241   The operation does not characterize a form but only the
        difference between forms.

5.242   The same operation which makes "q" from "p," makes
        "r" from "q," and so on. This can only be expressed by
        the fact that "p," "q," "r," etc., are variables which give
        general expression to certain formal relations.

5.25   The occurrence of an operation does not characterize
        the sense of a proposition.

        For an operation does not assert anything; only its
        result does, and this depends on the bases of the opera-
        tion.

        (Operation and function must not be confused with
        one another.)

5.251   A function cannot be its own argument, but the result
        of an operation can be its own basis.

5.252   Only in this way is the progress from term to term in a
        formal series possible (from type to type in the hier-
        archy of Russell and Whitehead). (Russell and White-
        head have not admitted the possibility of this progress
        but have made use of it all the same.)

5.2521　The repeated application of an operation to its own result I call its successive application ("$O' O'O' a$" is the result of the threefold successive application of "$O' \xi$" to "a").

In a similar sense I speak of the successive application of *several* operations to a number of propositions.

5.2522　The general term of the formal series a, $O' a$, $O' O'$ a, ... I write thus: "$[a, x, O' x]$." This expression in brackets is a variable. The first term of the expression is the beginning of the formal series, the second the form of an arbitrary term x of the series, and the third the form of that term of the series which immediately follows x.

5.2523　The concept of the successive application of an operation is equivalent to the concept "and so on."

5.253　One operation can reverse the effect of another. Operations can cancel one another.

5.254　Operations can vanish (e.g., denial in "$\sim \sim p$." $\sim \sim p = p$).

5.3　All propositions are results of truth-operations on the elementary propositions.

The truth-operation is the way in which a truth-function arises from elementary propositions.

According to the nature of truth-operations, in the same way as out of elementary propositions arise their truth-functions, from truth-functions arises a new one. Every truth-operation creates from truth-functions of elementary propositions another truth-function of elementary propositions, i.e., *a* proposition. The result of every truth-operation on the results of truth-operations on elementary propositions is also the result of *one* truth-operation on elementary propositions.

Every proposition is the result of truth-operations on elementary propositions.

5.31   The Schemata No. 4.31 are also significant, if "p," "q," "r," etc., are not elementary propositions.

And it is easy to see that the propositional sign in No. 4.42 expresses one truth-function of elementary propositions even when "p" and "q" are truth-functions of elementary propositions.

5.32   All truth-functions are results of the successive application of a finite number of truth-operations to elementary propositions.

5.4   Here it becomes clear that there are no such things as "logical objects" or "logical constants" (in the sense of Frege and Russell).

5.41   For all those results of truth-operations on truth-functions are identical, which are one and the same truth-function of elementary propositions.

5.42   That v, ⊃, etc., are not relations in the sense of right and left, etc., is obvious.

The possibility of crosswise definition of the logical "primitive signs" of Frege and Russell shows by itself that these are not primitive signs and that they signify no relations.

And it is obvious that the "⊃" which we define by means of "∼" and "v" is identical with that by which we define "v" with the help of "∼," and that this "v" is the same as the first, and so on.

5.43   That from a fact p an infinite number of *others* should follow, namely ∼ ∼p, ∼ ∼ ∼ ∼p, etc., is indeed hardly to be believed, and it is no less wonderful that the infinite number of propositions of logic (of mathematics) should follow from half a dozen "primitive propositions."

But all propositions of logic say the same thing. That is, nothing.

5.44   Truth-functions are not material functions.

If e.g., an affirmation can be produced by repeated

denial, is the denial—in any sense—contained in the affirmation?

Does "~ ~p" deny ~p, or does it affirm p; or both?

The proposition "~ ~p" does not treat of denial as an object, but the possibility of denial is already prejudged in affirmation.

And if there was an object called "~," then "~ ~p" would have to say something other than "p." For the one proposition would then treat of ~, the other would not.

5.441   This disappearance of the apparent logical constants also occurs if "~ (∃x). ~ fx" says the same as "(x). fx," or "(∃x). fx. x = a" the same as "fa."

5.442   If a proposition is given to us then the results of all truth-operations which have it as their basis are given *with* it.

5.45   If there are logical primitive signs a correct logic must make clear their position relative to one another and justify their existence. The construction of logic *out of* its primitive signs must become clear.

5.451   If logic has primitive ideas these must be independent of one another. If a primitive idea is introduced it must be introduced in all contexts in which it occurs at all. One cannot therefore introduce it for one context and then again for another. For example, if denial is introduced, we must understand it in propositions of the form "~p," just as in propositions like "~ (p v q)," "(∃x). ~ fx" and others. We may not first introduce it for one class of cases and then for another, for it would then remain doubtful whether its meaning in the two cases was the same, and there would be no reason to use the same way of symbolizing in the two cases.

(In short, what Frege ["Grundgesetze der Arithme-

tik"] has said about the introduction of signs by definitions holds, mutatis mutandis, for the introduction of
primitive signs also.)

5.452   The introduction of a new expedient in the symbolism
of logic must always be an event full of consequences.
No new symbol may be introduced in logic in brackets
or in the margin—with, so to speak, an entirely inno-
cent face.

(Thus in the "Principia Mathematica" of Russell and
Whitehead there occur definitions and primitive propositions in words. Why suddenly words here? This
would need a justification. There was none, and can be
none for the process is actually not allowed.)

But if the introduction of a new expedient has
proved necessary in one place, we must immediately
ask: Where is this expedient *always* to be used? Its position in logic must be made clear.

5.453   All numbers in logic must be capable of justification.

Or rather it must become plain that there are no
numbers in logic.

There are no preeminent numbers.

5.454   In logic there is no side by side, there can be no classification.

In logic there cannot be a more general and a more
special.

5.4541   The solution of logical problems must be simple for they
set the standard of simplicity.

Men have always thought that there must be a sphere
of questions whose answers—a priori—are symmetrical and united into a closed regular structure.

A sphere in which the proposition, simplex sigillum
veri, is valid.

5.46   When we have rightly introduced the logical signs, the sense of all their combinations has been already introduced with them: therefore not only "p v q" but also "$\sim(p v \sim q)$," etc. etc. We should then already have introduced the effect of all possible combinations of brackets; and it would then have become clear that the proper general primitive signs are not "p v q," "$(\exists x). fx$," etc., but the most general form of their combinations.

5.461　The apparently unimportant fact that the apparent relations like v and $\supset$ need brackets—unlike real relations is of great importance.

The use of brackets with these apparent primitive signs shows that these are not the real primitive signs; and nobody of course would believe that the brackets have meaning by themselves.

5.4611　Logical operation signs are punctuations.

5.47　It is clear that everything which can be said *beforehand* about the form of *all* propositions at all can be said *on one occasion*.

For all logical operations are already contained in the elementary proposition. For "fa" says the same as "$(\exists x). fx. x = a$."

Where there is composition, there is argument and function, and where these are, all logical constants already are.

One could say: the one logical constant is that which *all* propositions, according to their nature, have in common with one another.

That however is the general form of proposition.

5.471　The general form of proposition is the essence of proposition.

5.4711　To give the essence of proposition means to give the essence of all description, therefore the essence of the world.

5.472　The description of the most general propositional form is the description of the one and only general primitive sign in logic.

5.473   Logic must take care of itself.

A *possible* sign must also be able to signify. Everything which is possible in logic is also permitted. ("Socrates is identical" means nothing because there is no property which is called "identical." The proposition is senseless because we have not made some arbitrary determination, not because the symbol is in itself unpermissible.)

In a certain sense we cannot make mistakes in logic.

5.4731   Self-evidence, of which Russell has said so much, can only be discarded in logic by language itself preventing every logical mistake. That logic is a priori consists in the fact that we *cannot* think illogically.

5.4732   We cannot give a sign the wrong sense.

5.47321 Occam's razor is, of course, not an arbitrary rule nor one justified by its practical success. It simply says that *unnecessary* elements in a symbolism mean nothing.

Signs which serve *one* purpose are logically equivalent, signs which serve *no* purpose are logically meaningless.

5.4733   Frege says: Every legitimately constructed proposition must have a sense; and I say: Every possible proposition is legitimately constructed, and if it has no sense this can only be because we have given no *meaning* to some of its constituent parts.

(Even if we believe that we have done so.)

Thus "Socrates is identical" says nothing, because we have given *no* meaning to the word "identical" as *adjective*. For when it occurs as the sign of equality it symbolizes in an entirely different way—the symbolizing relation is another—therefore the symbol is in the two cases entirely different; the two symbols have the sign in common with one another only by accident.

5.474 The number of necessary fundamental operations depends *only* on our notation.

5.475 It is only a question of constructing a system of signs of a definite number of dimensions—of a definite mathematical multiplicity.

5.476 It is clear that we are not concerned here with a *number of primitive ideas* which must be signified but with the expression of a rule.

5.5 Every truth-function is a result of the successive application of the operation (-----T) (ξ, ...) to elementary propositions.

This operation denies all the propositions in the right-hand bracket and I call it the negation of these propositions.

5.501 An expression in brackets whose terms are propositions I indicate—if the order of the terms in the bracket is indifferent—by a sign of the form "($\bar{\xi}$)." "ξ" is a variable whose values are the terms of the expression in brackets, and the line over the variable indicates that it stands for all its values in the bracket.

(Thus if ξ has the 3 values P, Q, R, then($\bar{\xi}$) = [P, Q, R].)

The values of the variables must be determined.

The determination is the description of the propositions which the variable stands for.

How the description of the terms of the expression in brackets takes place is unessential.

We may distinguish 3 kinds of description: 1. Direct enumeration. In this case we can place simply its constant values instead of the variable. 2. Giving a function *fx*, whose values for all values of *x* are the propositions to be described. 3. Giving a formal law, according to which those propositions are constructed. In this case the terms of the expression in brackets are all the terms of a formal series.

5.502   Therefore I write instead of "(-----T) (ξ, ...)," "N ($\bar{\xi}$)."

N ($\bar{\xi}$) is the negation of all the values of the propositional variable ξ.

5.503   As it is obviously easy to express how propositions can be constructed by means of this operation and how propositions are not to be constructed by means of it, this must be capable of exact expression.

5.51   If ξ has only one value, then N($\bar{\xi}$) = $\sim p$ (not p), if it has two values then N($\bar{\xi}$) = $\sim p . \sim q$ (neither p nor q).

5.511   How can the all-embracing logic which mirrors the world use such special catches and manipulations? Only because all these are connected into an infinitely fine network, to the great mirror.

5.512   "$\sim p$" is true if "p" is false. Therefore in the true proposition "$\sim p$" "p" is a false proposition. How then can the stroke "$\sim$" bring it into agreement with reality?

That which denies in "$\sim p$" is however not "$\sim$," but that which all signs of this notation, which deny p, have in common.

Hence the common rule according to which "$\sim p$," "$\sim \sim \sim p$," "$\sim p \vee \sim p$," "$\sim p . \sim p$," etc., etc., (to infinity) are constructed. And this which is common to them all mirrors denial.

5.513   We could say: What is common to all symbols, which assert both p and q, is the proposition "$p . q$." What is common to all symbols, which assert either p or q, is the proposition "$p \vee q$."

And similarly we can say: Two propositions are opposed to one another when they have nothing in common with one another; and every proposition has only one negative, because there is only one proposition which lies altogether outside it.

Thus even in Russell's notation it is evident that "$q : pv$" says the same as "q"; that "$pv \sim p$" says nothing.

5.514　If a notation is fixed, there is in it a rule according to which all the propositions denying p are constructed, a rule according to which all the propositions asserting p are constructed, a rule according to which all the propositions asserting p or q are constructed, and so on. These rules are equivalent to the symbols and in them their sense is mirrored.

5.515　It must be recognized in our symbols that what is connected by "v," ".," etc., must be propositions.

And this is the case, for the symbols "p" and "q" presuppose "v," "$\sim$," etc. If the sign "p" in "$p \vee q$" does not stand for a complex sign, then by itself it cannot have sense; but then also the signs "$p \vee p$," "$p. p$," etc., which have the same sense as "p" have no sense. If, however, "$p \vee p$" has no sense, then also "$p \vee q$" can have no sense.

5.5151　Must the sign of the negative proposition be constructed by means of the sign of the positive? Why should one not be able to express the negative proposition by means of a negative fact? (Like: if "a" does not stand in a certain relation to "b," it could express that aRb is not the case.)

But here also the negative proposition is indirectly constructed with the positive.

The positive *proposition* must presuppose the existence of the negative *proposition* and conversely.

5.52　If the values of ξ are the total values of a function fx for all values x, then N $(\bar{\xi}) = \sim (\exists x). fx$.

5.521　I separate the concept *all* from the truth-function.

Frege and Russell have introduced generality in connexion with the logical product or the logical sum. Then it would be difficult to understand the propositions "$(\exists x) . fx$" and "$(x) . fx$" in which both ideas lie concealed.

5.522   That which is peculiar to the "symbolism of generality" is firstly, that it refers to a logical prototype, and secondly, that it makes constants prominent.

5.523   The generality symbol occurs as an argument.

5.524   If the objects are given, therewith are *all* objects also given.

If the elementary propositions are given, then therewith *all* elementary propositions are also given.

5.525   It is not correct to render the proposition "$(\exists x).fx$"—as Russell does—in words "fx is *possible*."

Certainty, possibility or impossibility of a state of affairs are not expressed by a proposition but by the fact that an expression is a tautology, a significant proposition or a contradiction.

That precedent to which one would always appeal, must be present in the symbol itself.

5.526   One can describe the world completely by completely generalized propositions, i.e., without from the outset coordinating any name with a definite object.

In order then to arrive at the customary way of expression we need simply say after an expression "there is one and only one x, which. . . .": and this x is a.

5.5261 A completely generalized proposition is like every other proposition composite. (This is shown by the fact that in "$[\exists x, \phi].\phi x$" we must mention "ϕ" and "x" separately. Both stand independently in signifying relations to the world as in the ungeneralized proposition.)

A characteristic of a composite symbol: it has something in common with *other* symbols.

5.5262   The truth or falsehood of *every* proposition alters something in the general structure of the world. And the range which is allowed to its structure by the totality of elementary propositions is exactly that which the completely general propositions delimit.

(If an elementary proposition is true, then, at any rate, there is one *more* elementary proposition true.)

5.53   Identity of the object I express by identity of the sign and not by means of a sign of identity. Difference of the objects by difference of the signs.

5.5301   That identity is not a relation between objects is obvious. This becomes very clear if, for example, one considers the proposition "$(x) : fx . \supset . x = a$." What this proposition says is simply that *only a* satisfies the function f, and not that only such things satisfy the function f which have a certain relation to a.

One could of course say that in fact *only a* has this relation to a, but in order to express this we should need the sign of identity itself.

5.5302   Russell's definition of "$=$" won't do; because according to it one cannot say that two objects have all their properties in common. (Even if this proposition is never true, it is nevertheless *significant*.)

5.5303   Roughly speaking: to say of *two* things that they are identical is nonsense, and to say of *one* thing that it is identical with itself is to say nothing.

5.531   I write therefore not "$f(a, b) . a = b$," but "$f(a, a)$" (or "$f(b, b)$"). And not "$f(a, b) . \sim a = b$," but "$f(a, b)$."

5.532   And analogously: not "$(\exists x, y) . f(x, y) . x = y$," but "$(\exists x) . f(x, x)$"; and not "$(\exists x, y) . f(x, y) . \sim x = y$," but "$(\exists x, y) . f(x, y)$."

(Therefore instead of Russell's "$(\exists x, y) . f(x, y)$" : "$(\exists x, y). f(x, y) . v. (\exists x) . f(x, x).$")

5.5321   Instead of "$(x) : fx \supset x = a$" we therefore write e.g., "$(\exists x) . fx . \supset . fa : \sim (\exists x, y). fx . fy.$"

And the proposition "*only* one x satisfies $f(\ )$" reads; "$(\exists x) . fx : \sim (\exists x, y) . fx . fy.$"

5.533   The identity sign is therefore not an essential constituent of logical notation.

5.534   And we see that apparent propositions like: "$a = a$," "$a = b . b = c . \supset a = c$," "$(x) . x = x$," "$(\exists x) . x = a$," etc. cannot be written in a correct logical notation at all.

5.535   So all problems disappear which are connected with such pseudo-propositions.

This is the place to solve all the problems which arise through Russell's "Axiom of Infinity."

What the axiom of infinity is meant to say would be expressed in language by the fact that there is an infinite number of names with different meanings.

5.5351   There are certain cases in which one is tempted to use expressions of the form "$a = a$" or "$p \supset p$" and of that kind. And indeed this takes place when one would like to speak of the archetype Proposition, Thing, etc. So Russell in the *Principles of Mathematics* has rendered the nonsense "p is a proposition" in symbols by "$p \supset p$" and has put it as hypothesis before certain propositions to show that their places for arguments could only be occupied by propositions.

(It is nonsense to place the hypothesis $p \supset p$ before a proposition in order to ensure that its arguments have the right form, because the hypothesis for a non-proposition as argument becomes not false but meaningless, and because the proposition itself becomes senseless for arguments of the wrong kind, and therefore it survives the wrong arguments no better and no worse than the senseless hypothesis attached for this purpose.)

5.5352   Similarly it was proposed to express "There are no things" by "$\sim (\exists x) \, . \, x = x$." But even if this were a proposition—would it not be true if indeed "There were things," but these were not identical with themselves?

5.54   In the general propositional form, propositions occur in a proposition only as bases of the truth-operations.

5.541   At first sight it appears as if there were also a different way in which one proposition could occur in another.

Especially in certain propositional forms of psychology, like "A thinks, that p is the case," or "A thinks p," etc.

Here it appears superficially as if the proposition p stood to the object A in a kind of relation.

(And in modern epistomology [Russell, Moore, etc.] those propositions have been conceived in this way.)

5.542   But it is clear that "A believes that p," "A thinks p," "A says p," are of the form "'p' says p": and here we have no coordination of a fact and an object, but a coordination of facts by means of a coordination of their objects.

5.5421   This shows that there is no such thing as the soul—the subject, etc.—as it is conceived in contemporary superficial psychology.

A composite soul would not be a soul any longer.

5.5422   The correct explanation of the form of the proposition "A judges p" must show that it is impossible to judge a nonsense. (Russell's theory does not satisfy this condition.)

5.5423   To perceive a complex means to perceive that its constituents are combined in such and such a way.

This perhaps explains that the figure

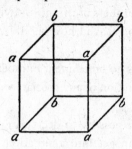

can be seen in two ways as a cube; and all similar phenomena. For we really see two different facts.

(If I fix my eyes first on the corners *a* and only glance at *b*, *a* appears in front and *b* behind, and vice versa.)

5.55   We must now answer a priori the question as to all possible forms of the elementary propositions.

The elementary proposition consists of names. Since we cannot give the number of names with different meanings, we cannot give the composition of the elementary proposition.

5.551   Our fundamental principle is that every question which can be decided at all by logic can be decided without further trouble.

(And if we get into a situation where we need to answer such a problem by looking at the world, this shows that we are on a fundamentally wrong track.)

5.552   The "experience" which we need to understand logic is not that such and such is the case, but that something *is;* but that is *no* experience.

Logic *precedes* every experience—that something is *so.*

It is before the How, not before the What.

5.5521   And if this were not the case, how could we apply logic? We could say: if there were a logic, even if there were no world, how then could there be a logic, since there is a world?

5.553   Russell said that there were simple relations between different numbers of things (individuals). But between what numbers? And how should this be decided—by experience?

(There is no preeminent number.)

5.554   The enumeration of any special forms would be entirely arbitrary.

5.5541 It should be possible to decide a priori whether, for example, I can get into a situation in which I need to symbolize with a sign of a 27-termed relation.

5.5542 May we then ask this at all? Can we set out a sign form and not know whether anything can correspond to it?

Has the question sense: what must *be* in order that something can be the case?

5.555   It is clear that we have a concept of the elementary proposition apart from its special logical form.

Where, however, we can build symbols according to a system, there this system is the logically important thing and not the single symbols.

And how would it be possible that I should have to deal with forms in logic which I can invent: but I must have to deal with that which makes it possible for me to invent them.

5.556   There cannot be a hierarchy of the forms of the elementary propositions. Only that which we ourselves construct can we foresee.

5.5561 Empirical reality is limited by the totality of objects. The boundary appears again in the totality of elementary propositions.

The hierarchies are and must be independent of reality.

5.5562 If we know on purely logical grounds, that there must be elementary propositions, then this must be known by everyone who understands the propositions in their unanalysed form.

5.5563   All propositions of our colloquial language are actually, just as they are, logically completely in order. That most simple thing which we ought to give here is not a simile of truth but the complete truth itself.

(Our problems are not abstract but perhaps the most concrete that there are.)

5.557   The *application* of logic decides what elementary propositions there are.

What lies in the application logic cannot anticipate.

It is clear that logic may not collide with its application.

But logic must have contact with its application.

Therefore logic and its application may not overlap one another.

5.5571   If I cannot give elementary propositions a priori then it must lead to obvious nonsense to try to give them.

5.6   *The limits of my language* mean the limits of my world.

5.61   Logic fills the world: the limits of the world are also its limits.

We cannot therefore say in logic: This and this there is in the world, that there is not.

For that would apparently presuppose that we exclude certain possibilities, and this cannot be the case since otherwise logic must get outside the limits of the world: that is, if it could consider these limits from the other side also.

What we cannot think, that we cannot think: we cannot therefore *say* what we cannot think.

5.62   This remark provides a key to the question, to what extent solipsism is a truth.

In fact what solipsism *means*, is quite correct, only it cannot be *said*, but it shows itself.

That the world is *my* world, shows itself in the fact that the limits of the language (the language which only I understand) mean the limits of *my* world.

5.621   The world and life are one.

5.63   I am my world. (The microcosm.)

5.631   The thinking, presenting subject; there is no such thing.

If I wrote a book "The world as I found it," I should also have therein to report on my body and say which members obey my will and which do not, etc. This then would be a method of isolating the subject or rather of showing that in an important sense there is no subject: that is to say, of it alone in this book mention could *not* be made.

5.632   The subject does not belong to the world but it is a limit of the world.

5.633   *Where in* the world is a metaphysical subject to be noted?

You say that this case is altogether like that of the eye and the field of sight. But you do *not* really see the eye.

And from nothing *in the field of sight* can it be concluded that it is seen from an eye.

5.6331   For the field of sight has not a form like this:

5.634   This is connected with the fact that no part of our experience is also a priori.

Everything we see could also be otherwise.

Everything we can describe at all could also be otherwise.

There is no order of things a priori.

5.64 Here we see that solipsism strictly carried out coincides with pure realism. The I in solipsism shrinks to an extensionless point and there remains the reality coordinated with it.

5.641 There is therefore really a sense in which in philosophy we can talk of a non-psychological I.

The I occurs in philosophy through the fact that the "world is my world."

The philosophical I is not the man, not the human body or the human soul of which psychology treats, but the metaphysical subject, the limit—not a part of the world.

6 The general form of truth-function is: $[\bar{p}, \bar{\xi}, N(\bar{\xi})]$
This is the general form of proposition.

6.001 This says nothing else than that every proposition is the result of successive applications of the operation $N'(\bar{\xi})$ to the elementary propositions.

6.002 If we are given the general form of the way in which a proposition is constructed, then thereby we are also given the general form of the way in which by an operation out of one proposition another can be created.

6.01 The general form of the operation $\Omega'(\bar{\eta})$. is therefore: $[\bar{\xi}, N(\bar{\xi})]'(\bar{\eta})(=[\bar{\eta}, \bar{\xi}, N(\bar{\xi})])$.

This is the most general form of transition from one proposition to another.

6.02 And thus we come to numbers: I define

$x = \Omega^0 x$ Def. and

$$\Omega'\Omega^{v'}x = \Omega^{v\varnothing1'}x \text{ Def.}$$

According, then, to these symbolic rules we write the series $x, \Omega'x, \Omega'\Omega'x, \Omega'\Omega'\Omega'x \ldots$

as: $\Omega^{0'}x, \Omega^{0+1'}x, \Omega^{0+1+1'}x, \Omega^{0+1+1+1'}x \ldots$

Therefore I write in place of "$[x, \xi, \Omega' \xi]$,"

"$[\Omega^{0'}x, \Omega^{v'}x, \Omega^{v\emptyset1'}x]$."

And I define:

$0 + 1 = 1$ Def.

$0 + 1 + 1 = 2$ Def.

$0 + 1 + 1 + 1 = 3$ Def.

and so on.

6.021   A number is the exponent of an operation.

6.022   The concept number is nothing else than that which is common to all numbers, the general form of number.

       The concept number is the variable number.

       And the concept of equality of numbers is the general form of all special equalities of numbers.

6.03   The general form of the cardinal number is: $[0, \xi, \xi + 1]$.

6.031   The theory of classes is altogether superfluous in mathematics.

       This is connected with the fact that the generality which we need in mathematics is not the *accidental* one.

6.1   The propositions of logic are tautologies.

6.11   The propositions of logic therefore say nothing. (They are the analytical propositions.)

6.111   Theories which make a proposition of logic appear substantial are always false. One could e.g., believe that the words "true" and "false" signify two properties among other properties, and then it would appear as a remarkable fact that every proposition possesses one of these properties. This now by no means appears self-evident, no more so than the proposition "All roses are either yellow or red" would sound even if it were true. Indeed our proposition now gets quite the character of a proposition of natural science and this is a certain symptom of its being falsely understood.

6.112 The correct explanation of logical propositions must give them a peculiar position among all propositions.

6.113 It is the characteristic mark of logical propositions that one can perceive in the symbol alone that they are true; and this fact contains in itself the whole philosophy of logic. And so also it is one of the most important facts that the truth or falsehood of non-logical propositions can *not* be recognized from the propositions alone.

6.12 The fact that the propositions of logic are tautologies *shows* the formal—logical—properties of language, of the world.

That its constituent parts connected together *in this way* give a tautology characterizes the logic of its constituent parts.

In order that propositions connected together in a definite way may give a tautology they must have definite properties of structure. That they give a tautology when *so* connected shows therefore that they possess these properties of structure.

6.1201 That e.g., the propositions "p" and "$\sim p$" in the connexion "$\sim (p . \sim p)$" give a tautology shows that they contradict one another. That the propositions "$p \supset q$," "p" and "q" connected together in the form "$(p \supset q) . (p) : \supset : (q)$" give a tautology shows that q follows from p and $p \supset q$. That "$(x) . fx : \supset : fa$" is a tautology shows that fa follows from $(x) . fx$, etc. etc.

6.1202 It is clear that we could have used for this purpose contradictions instead of tautologies.

6.1203 In order to recognize a tautology as such, we can, in cases in which no sign of generality occurs in the tautology, make use of the following intuitive method: I write instead of "p," "q," "r," etc., "TpF," "TqF," "Trf," etc. The truth-combinations I express by brackets, e.g.,:

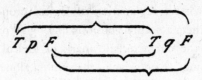

and the coordination of the truth or falsity of the whole proposition with the truth-combinations of the truth-arguments by lines in the following way:

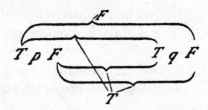

This sign, for example, would therefore present the proposition $p \supset q$. Now I will proceed to inquire whether such a proposition as $\sim (p \,.\sim p)$ (The Law of Contradiction) is a tautology. The form "$\sim \xi$" is written in our notation.

the form "$\xi . \eta$" thus:—

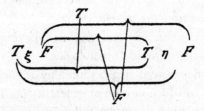

Hence the proposition $\sim (p. \sim q)$ runs thus:—

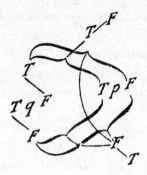

If here we put "*p*" instead of "*q*" and examine the combination of the outermost T and F with the innermost, it is seen that the truth of the whole proposition is coordinated with *all* the truth-combinations of its argument, its falsity with none of the truth-combinations.

6.121   The propositions of logic demonstrate the logical properties of propositions, by combining them into propositions which say nothing.

This method could be called a zero-method. In a logical proposition propositions are brought into equilibrium with one another, and the state of equilibrium then shows how these propositions must be logically constructed.

6.122   Whence it follows that we can get on without logical propositions, for we can recognize in an adequate notation the formal properties of the propositions by mere inspection.

6.1221   If for example two propositions "*p*" and "*q*" *give a* tautology in the connexion "$p \supset q$," then it is clear that *q* follows from *p*.

E.g., that "*q*" follows from "$p \supset q \,.\, p$" we see from these two propositions themselves, but we can also show it by combining them to "$p \supset q \,.\, p : \supset : q$" and then showing that this is a tautology.

6.1222   This throws light on the question why logical proposi-
         tions can no more be empirically established than they
         can be empirically refuted. Not only must a proposition
         of logic be incapable of being contradicted by any pos-
         sible experience, but it must also be incapable of being
         established by any such.

6.1223   It now becomes clear why we often feel as though "logi-
         cal truths" must be "*postulated*" by us. We can in fact
         postulate them in so far as we can postulate an adequate
         notation.

6.1224   It also becomes clear why logic has been called the the-
         ory of forms and of inference.

6.123   It is clear that the laws of logic cannot themselves obey
         further logical laws.

         (There is not, as Russell supposed, for every "type" a
         special law of contradiction; but one is sufficient, since
         it is not applied to itself.)

6.1231   The mark of logical propositions is not their general
         validity.

         To be general is only to be accidentally valid for all
         things. An ungeneralized proposition can be tautolo-
         gous just as well as a generalized one.

6.1232   Logical general validity, we could call essential as op-
         posed to accidental general validity, e.g., of the proposi-
         tion "all men are mortal." Propositions like Russell's
         "axiom of reducibility" are not logical propositions, and
         this explains our feeling that, if true, they can only be
         true by a happy chance.

6.1233   We can imagine a world in which the axiom of reduc-
         ibility is not valid. But it is clear that logic has nothing
         to do with the question whether our world is really of
         this kind or not.

6.124   The logical propositions describe the scaffolding of the
         world, or rather they present it. They "treat" of nothing.

They presuppose that names have meaning, and that elementary propositions have sense. And this is their connexion with the world. It is clear that it must show something about the world that certain combinations of symbols—which essentially have a definite character—are tautologies. Herein lies the decisive point. We said that in the symbols which we use much is arbitrary, much not. In logic only this expresses: but this means that in logic it is not *we* who express, by means of signs, what we want, but in logic the nature of the essentially necessary signs itself asserts. That is to say, if we know the logical syntax of any sign language, then all the propositions of logic are already given.

6.125   It is possible, even in the old logic, to give at the outset a description of all "true" logical propositions.

6.1251   Hence there can *never* be surprises in logic.

6.126   Whether a proposition belongs to logic can be determined by determining the logical properties of the *symbol*.

And this we do when we prove a logical proposition. For without troubling ourselves about a sense and a meaning, we form the logical propositions out of others by mere *symbolic rules*.

We prove a logical proposition by creating it out of other logical propositions by applying in succession certain operations, which again generate tautologies out of the first. (And from a tautology only tautologies *follow*.)

Naturally this way of showing that its propositions are tautologies is quite unessential to logic. Because the propositions, from which the proof starts, must show without proof that they are tautologies.

6.1261   In logic process and result are equivalent. (Therefore no surprises.)

6.1262   Proof in logic is only a mechanical expedient to facilitate the recognition of tautology, where it is complicated.

6.1263   It would be too remarkable, if one could prove a significant proposition *logically* from another, and a logical proposition *also*. It is clear from the beginning that the logical proof of a significant proposition and the proof *in* logic must be two quite different things.

6.1264   The significant proposition asserts something, and its proof shows that it is so; in logic every proposition is the form of a proof.

Every proposition of logic is a modus ponens presented in signs. (And the modus ponens cannot be expressed by a proposition.)

6.1265   Logic can always be conceived to be such that every proposition is its own proof.

6.127   All propositions of logic are of equal rank; there are not some which are essentially primitive and others deduced from these.

Every tautology itself shows that it is a tautology.

6.1271   It is clear that the number of "primitive propositions of logic" is arbitrary, for we could deduce logic from one primitive proposition by simply forming, for example, the logical product of Frege's primitive propositions. (Frege would perhaps say that this would no longer be immediately self-evident. But it is remarkable that so exact a thinker as Frege should have appealed to the degree of self-evidence as the criterion of a logical proposition.)

6.13   Logic is not a theory but a reflexion of the world.

Logic is transcendental.

6.2   Mathematics is a logical method.

The propositions of mathematics are equations, and therefore pseudo-propositions.

6.21   Mathematical propositions express no thoughts.

6.211   In life it is never a mathematical proposition which we need, but we use mathematical propositions *only* in order to infer from propositions which do not belong to mathematics to others which equally do not belong to mathematics.

(In philosophy the question "Why do we really use that word, that proposition?" constantly leads to valuable results.)

6.22   The logic of the world which the propositions of logic show in tautologies, mathematics shows in equations.

6.23   If two expressions are connected by the sign of equality, this means that they can be substituted for one another. But whether this is the case must show itself in the two expressions themselves.

It characterizes the logical form of two expressions, that they can be substituted for one another.

6.231   It is a property of affirmation that it can be conceived as double denial.

It is a property of "$1 + 1 + 1 + 1$" that it can be conceived as "$(1 + 1) + (1 + 1)$."

6.232   Frege says that these expressions have the same meaning but different senses.

But what is essential about equation is that it is not necessary in order to show that both expressions, which are connected by the sign of equality, have the same meaning: for this can be perceived from the two expressions themselves.

6.2321   And, that the propositions of mathematics can be proved means nothing else than that their correctness can be seen without our having to compare what they express with the facts as regards correctness.

6.2322   The identity of the meaning of two expressions cannot be *asserted*. For in order to be able to assert anything about their meaning, I must know their meaning, and if I know their meaning, I know whether they mean the same or something different.

6.2323   The equation characterizes only the standpoint from which I consider the two expressions, that is to say the standpoint of their equality of meaning.

6.233   To the question whether we need intuition for the solution of mathematical problems it must be answered that language itself here supplies the necessary intuition.

6.2331   The process of calculation brings about just this intuition.

Calculation is not an experiment.

6.234   Mathematics is a method of logic.

6.2341   The essential of mathematical method is working with equations. On this method depends the fact that every proposition of mathematics must be self-intelligible.

6.24   The method by which mathematics arrives at its equations is the method of substitution.

For equations express the substitutability of two expressions, and we proceed from a number of equations to new equations, replacing expressions by others in accordance with the equations.

6.241   Thus the proof of the proposition $2 \times 2 = 4$ runs:

$$(\Omega^v)^\mu x = \Omega^{v \times \mu} x \text{ Def.}$$

$$\Omega^{2 \times 2}{}'x = (\Omega^2)^{2}{}'x = (\Omega^2)^{1+1}{}'x = \Omega^{2}{}'\Omega^{2}{}'x = \Omega^{1+1}{}'\Omega^{1+1}{}'x$$
$$= (\Omega'\ \Omega)'(\Omega'\ \Omega)'x = \Omega'\Omega'\Omega'\Omega'x = \Omega^{1+1+1+1}{}'x = \Omega^{4}{}'x.$$

6.3   Logical research means the investigation of *all regularity*. And outside logic all is accident.

6.31   The so-called law of induction cannot in any case be a logical law, for it is obviously a significant proposition.— And therefore it cannot be a law a priori either.

6.32   The law of causality is not a law but the form of a law.*

6.321   "Law of Causality" is a class name. And as in mechanics there are, for instance, minimum-laws, such as that

---

* I.e., not the form of one particular law, but of any law of a certain sort (B.R.)

of least action, so in physics there are causal laws, laws of the causality form.

6.3211   Men had indeed an idea that there must be *a* "law of least action," before they knew exactly how it ran. (Here, as always, the a priori certain proves to be something purely logical.)

6.33   We do not *believe* a priori in a law of conservation, but we *know* a priori the possibility of a logical form.

6.34   All propositions, such as the law of causation, the law of continuity in nature, the law of least expenditure in nature, etc. etc., all these are a priori intuitions of possible forms of the propositions of science.

6.341   Newtonian mechanics, for example, brings the description of the universe to a unified form. Let us imagine a white surface with irregular black spots. We now say: Whatever kind of picture these make I can always get as near as I like to its description, if I cover the surface with a sufficiently fine square network and now say of every square that it is white or black. In this way I shall have brought the description of the surface to a unified form. This form is arbitrary, because I could have applied with equal success a net with a triangular or hexagonal mesh. It can happen that the description would have been simpler with the aid of a triangular mesh; that is to say we might have described the surface more accurately with a triangular, and coarser, than with the finer square mesh, or vice versa, and so on. To the different networks correspond different systems of describing the world. Mechanics determine a form of description by saying: All propositions in the description of the world must be obtained in a given way from a number of given propositions—the mechanical axioms. It thus provides the bricks for building the edifice of science, and says: Whatever building thou wouldst erect, thou shalt construct it in some manner with these bricks and these alone.

(As with the system of numbers one must be able to write down any arbitrary number, so with the system of mechanics one must be able to write down any arbitrary physical proposition.)

6.342 And now we see the relative position of logic and mechanics. (We could construct the network out of figures of different kinds, as out of triangles and hexagons together.) That a picture like that instanced above can be described by a network of a given form asserts *nothing* about the picture. (For this holds of every picture of this kind.) But *this* does characterize the picture, the fact, namely, that it can be *completely* described by a definite net of *definite* fineness.

So too the fact that it can be described by Newtonian mechanics asserts nothing about the world; but *this* asserts something, namely, that it can be described in that particular way in which it is described, as is indeed the case. The fact, too, that it can be described more simply by one system of mechanics than by another says something about the world.

6.343 Mechanics is an attempt to construct according to a single plan all *true* propositions which we need for the description of the world.

6.3431 Through the whole apparatus of logic the physical laws still speak of the objects of the world.

6.3432 We must not forget that the description of the world by mechanics is always quite general. There is, for example, never any mention of *particular* material points in it, but always only of *some points or other*.

6.35 Although the spots in our picture are geometrical figures, geometry can obviously say nothing about their actual form and position. But the network is *purely* geometrical, and all its properties can be given a priori.

Laws, like the law of causation, etc., treat of the network and not of what the network described.

6.36   If there were a law of causality, it might run: "There are natural laws."

But that can clearly not be said: it shows itself.

6.361   In the terminology of Hertz we might say: Only *uniform* connexions are *thinkable*.

6.3611   We cannot compare any process with the "passage of time"—there is no such thing—but only with another process (say, with the movement of the chronometer).

Hence the description of the temporal sequence of events is only possible if we support ourselves on another process.

It is exactly analogous for space. When, for example, we say that neither of two events (which mutually exclude one another) can occur, because there is *no cause* why the one should occur rather than the other, it is really a matter of our being unable to describe *one* of the two events unless there is some sort of asymmetry. And if there *is* such an asymmetry, we can regard this as the *cause* of the occurrence of the one and of the non-occurrence of the other.

6.36111   The Kantian problem of the right and left hand which cannot be made to cover one another already exists in the plane, and even in one-dimensional space; where the two congruent figures *a* and *b* cannot be made to cover one another without

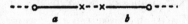

moving them out of this space. The right and left hand are in fact completely congruent. And the fact that they cannot be made to cover one another has nothing to do with it.

A right-hand glove could be put on a left hand if it could be turned round in four-dimensional space.

6.362 What can be described can happen too, and what is excluded by the law of causality cannot be described.

6.363 The process of induction is the process of assuming the *simplest* law that can be made to harmonize with our experience.

6.3631 This process, however, has no logical foundation but only a psychological one.

It is clear that there are no grounds for believing that the simplest course of events will really happen.

6.36311 That the sun will rise tomorrow, is an hypothesis; and that means that we do not *know* whether it will rise.

6.37 A necessity for one thing to happen because another has happened does not exist. There is only *logical* necessity.

6.371 At the basis of the whole modern view of the world lies the illusion that the so-called laws of nature are the explanations of natural phenomena.

6.372 So people stop short at natural laws as at something unassailable, as did the ancients at God and Fate.

And they both are right and wrong. But the ancients were clearer, in so far as they recognized one clear conclusion, whereas in the modern system it should appear as though *everything* were explained.

6.373 The world is independent of my will.

6.374 Even if everything we wished were to happen, this would only be, so to speak, a favour of fate, for there is no *logical* connexion between will and world, which would guarantee this, and the assumed physical connexion itself we could not again will.

6.375 As there is only a *logical* necessity, so there is only a *logical* impossibility.

6.3751 For two colours, e.g., to be at one place in the visual field, is impossible, logically impossible, for it is excluded by the logical structure of colour.

Let us consider how this contradiction presents itself in physics. Somewhat as follows: That a particle cannot at the same time have two velocities, i.e., that at the same time it cannot be in two places, i.e., that particles in different places at the same time cannot be identical.

(It is clear that the logical product of two elementary propositions can neither be a tautology nor a contradiction. The assertion that a point in the visual field has two different colours at the same time, is a contradiction.)

6.4   All propositions are of equal value.

6.41   The sense of the world must lie outside the world. In the world everything is as it is and happens as it does happen. *In* it there is no value—and if there were, it would be of no value.

If there is a value which is of value, it must lie outside all happening and being-so. For all happening and being-so is accidental.

What makes it non-accidental cannot lie *in* the world, for otherwise this would again be accidental.

It must lie outside the world.

6.42   Hence also there can be no ethical propositions. Propositions cannot express anything higher.

6.421   It is clear that ethics cannot be expressed.

Ethics are transcendental.

(Ethics and æsthetics are one.)

6.422   The first thought in setting up an ethical law of the form "thou shalt . . ." is: And what if I do not do it. But it is clear that ethics has nothing to do with punishment and reward in the ordinary sense. This question as to the *consequences* of an action must therefore be irrelevant. At least these consequences will not be events.

For there must be something right in that formulation of the question. There must be some sort of ethical reward and ethical punishment, but this must lie in the action itself.

(And this is clear also that the reward must be something acceptable, and the punishment something unacceptable.)

6.423   Of the will as the bearer of the ethical we cannot speak.

And the will as a phenomenon is only of interest to psychology.

6.43   If good or bad willing changes the world, it can only change the limits of the world, not the facts; not the things that can be expressed in language.

In brief, the world must thereby become quite another. It must so to speak wax or wane as a whole.

The world of the happy is quite another than that of the unhappy.

6.431   As in death, too, the world does not change, but ceases.

6.4311 Death is not an event of life. Death is not lived through.

If by eternity is understood not endless temporal duration but timelessness, then he lives eternally who lives in the present.

Our life is endless in the way that our visual field is without limit.

6.4312 The temporal immortality of the soul of man, that is to say, its eternal survival also after death, is not only in no way guaranteed, but this assumption in the first place will not do for us what we always tried to make it do. Is a riddle solved by the fact that I survive forever? Is this eternal life not as enigmatic as our present one? The solution of the riddle of life in space and time lies *outside* space and time.

(It is not problems of natural science which have to be solved.)

6.432   *How* the world is, is completely indifferent for what is higher. God does not reveal himself *in* the world.

6.4321   The facts all belong only to the task and not to its performance.

6.44   Not *how* the world is, is the mystical, but *that* it is.

6.45   The contemplation of the world sub specie aeterni is its contemplation as a limited whole.

        The feeling of the world as a limited whole is the mystical feeling.

6.5   For an answer which cannot be expressed the question too cannot be expressed.

        The riddle does not exist.

        If a question can be put at all, then it *can* also be answered.

6.51   Scepticism is *not* irrefutable, but palpably senseless, if it would doubt where a question cannot be asked.

        For doubt can only exist where there is a question; a question only where there is an answer, and this only where something *can* be *said*.

6.52   We feel that even if *all possible* scientific questions be answered, the problems of life have still not been touched at all. Of course there is then no question left, and just this is the answer.

6.521   The solution of the problem of life is seen in the vanishing of this problem.

        (Is not this the reason why men to whom after long doubting the sense of life became clear, could not then say wherein this sense consisted?)

6.522   There is indeed the inexpressible. This *shows* itself; it is the mystical.

6.53   The right method of philosophy would be this. To say nothing except what can be said, i.e., the propositions of natural science, i.e., something that has nothing to do with philosophy: and then always, when someone else wished to say something metaphysical, to demonstrate to him that he had given no meaning to certain signs in his propositions. This method would be unsatisfying to the other—he would not have the feeling that we were teaching him philosophy—but it would be the only strictly correct method.

6.54   My propositions are elucidatory in this way: he who understands me finally recognizes them as senseless, when he has climbed out through them, on them, over them. (He must so to speak throw away the ladder, after he has climbed up on it.)

He must surmount these propositions; then he sees the world rightly.

7   Whereof one cannot speak, thereof one must be silent.

STUDIES FOR 'PHILOSOPHICAL INVESTIGATIONS'

THE BLUE BOOK

Let us attack this question by asking, first, what is an explanation of the meaning of a word; what does the explanation of a word look like?

The way this question helps us is analogous to the way the question "how do we measure a length?" helps us to understand the problem "what is length?"

The questions "What is length?", "What is meaning?", "What is the number one?" etc., produce in us a mental cramp. We feel that we can't point to anything in reply to them and yet ought to point to something. (We are up against one of the great sources of philosophical bewilderment: a substantive makes us look for a thing that corresponds to it.)

Asking first "What's an explanation of meaning?" has two advantages. You in a sense bring the question "What is meaning?" down to earth. For, surely, to understand the meaning of "meaning" you ought also to understand the meaning of "explanation of meaning". Roughly: "let's ask what the explanation of meaning is, for whatever that explains will be the meaning." Studying the grammar of the expression "explanation of meaning" will teach you something about the grammar of the word "meaning" and will cure you of the temptation to look about you for some object which you might call "the meaning".

What one generally calls "explanations of the meaning of a word" can, *very roughly*, be divided into verbal and ostensive definitions. It will be seen later in what sense this division is only rough and provisional (and that it is, is an important point). The verbal definition, as it takes us from one verbal expression to another, in a sense gets us no further. In the ostensive definition however we seem to make a much more real step towards learning the meaning.

One difficulty which strikes us is that for many words in our language there do not seem to be ostensive definitions; e.g. for such words as "one", "number", "not", etc.

Question: Need the ostensive definition itself be understood?—Can't the ostensive definition be misunderstood?

If the definition explains the meaning of a word, surely it can't be essential that you should have heard the word before. It is the ostensive definition's business to *give* it a meaning. Let us then explain the word "tove" by pointing to a pencil and saying "this is tove". (Instead of "this is tove" I could here have said "this is called 'tove'". I point this out to remove, once and for all, the idea that the words of the ostensive definition predicate something of the defined; the confusion between the sentence "this is red", attributing the colour red to something, and the ostensive definition "this is called 'red'".) Now the ostensive definition "this is tove" can be interpreted in all sorts of ways. I will give a few such interpretations and use English words with well established usage. The definition then can be interpreted to mean:

"This is a pencil",
"This is round",
"This is wood",
"This is one",
"This is hard", etc. etc.

One might object to this argument that all these interpretations presuppose another word-language. And this objection is significant if by "interpretation" we only mean "translation into a word-language".—Let me give some hints which might make this clearer. Let us ask ourselves what is our criterion when we say that someone has interpreted the ostensive definition in a particular way. Suppose I give to an Englishman the ostensive definition "this is what the Germans call 'Buch'". Then, in the great majority of cases at any rate, the English word "book" will come into the Englishman's mind. We may say he has interpreted "Buch" to mean "book". The case will be different if e.g. we point to a thing which he has never seen before and say: "This is a banjo". Possibly the word "guitar" will then come into his mind, possibly no word at all but the image of a similar instrument, possibly nothing at all. Supposing then I give him the order "now pick a banjo from amongst these things." If he picks what we call a "banjo" we might say "he has given the word 'banjo' the correct interpretation"; if he picks some other instrument— "he has interpreted 'banjo' to mean 'string instrument'".

We say "he has given the word 'banjo' this or that interpretation", and are inclined to assume a definite act of interpretation besides the act of choosing.

Our problem is analogous to the following:

If I give someone the order "fetch me a red flower from that meadow", how is he to know what sort of flower to bring, as I have only given him a *word*?

Now the answer one might suggest first is that he went to look for a red flower carrying a red image in his mind, and comparing it with the flowers to see which of them had the colour of the image. Now there is such a way of searching, and it is not at all essential that the image we use should be a mental one. In fact the process may be this: I carry a chart co-ordinating names and coloured squares. When I hear the order "fetch me etc." I draw my finger across the chart from the word "red" to a certain square, and I go and look for a flower which has the same colour as the square. But this is not the only way of searching and it isn't the usual way. We go, look about us, walk up to a flower and pick it, without comparing it to anything. To see that the process of obeying the order can be of this kind, consider the order "*imagine* a red patch". You are not tempted in this case to think that *before* obeying you must have imagined a red patch to serve you as a pattern for the red patch which you were ordered to imagine.

Now you might ask: do we *interpret* the words before we obey the order? And in some cases you will find that you do something which might be called interpreting before obeying, in some cases not.

It seems that there are *certain definite* mental processes bound up with the working of language, processes through which alone language can function. I mean the processes of understanding and meaning. The signs of our language seem dead without these mental processes; and it might seem that the only function of the signs is to induce such processes, and that these are the things we ought really to be interested in. Thus, if you are asked what is the relation between a name and the thing it names, you will be inclined to answer that the relation is a psychological one, and perhaps when you say this you think in particular of the mechanism of association.—We are tempted to think that the action of language consists of two parts: an inorganic part, the handling of signs, and an organic part, which we may call understanding these signs, meaning them, interpreting them, thinking. These latter activities seem to take place in a queer kind of medium, the mind; and the mechanism of the mind, the nature of which, it seems, we don't quite understand, can bring about effects which no material mechanism could. Thus e.g. a thought (which is such a mental process) can agree or disagree with reality; I am able to think of a man who isn't present; I am able to imagine him, 'mean him' in a remark which I make about him, even if he is thousands of miles away or dead. "What a queer mechanism," one might say, "the mechanism of wishing must be if I can wish that which will never happen".

There is one way of avoiding at least partly the occult appearance of the processes of thinking, and it is, to replace in these processes any working of the imagination by acts of looking at real objects. Thus it may seem essential that, at least in certain cases, when I hear the word "red" with understanding, a red image should be before my mind's eye. But why should I not substitute seeing a red bit of paper for imagining a red patch? The visual image will only be the more vivid. Imagine a man always carrying a sheet of paper in his pocket on which the names of colours are co-ordinated with coloured patches. You may say that it would be a nuisance to carry such a table of samples about with you, and that the mechanism of association is what we always use instead of it. But this is irrelevant; and in many cases it is not even true. If, for instance, you were ordered to paint a particular shade of blue called "Prussian Blue", you might have to use a table to lead you from the word "Prussian Blue" to a sample of the colour, which would serve you as your copy.

We could perfectly well, for our purposes, replace every process of imagining by a process of looking at an object or by painting, drawing or modelling; and every process of speaking to oneself by speaking aloud or by writing.

Frege ridiculed the formalist conception of mathematics by saying that the formalists confused the unimportant thing, the sign, with the important, the meaning. Surely, one wishes to say, mathematics does not treat of dashes on a bit of paper. Frege's idea could be expressed thus: the propositions of mathematics, if they were just complexes of dashes, would be dead and utterly uninteresting, whereas they obviously have a kind of life. And the same, of course, could be said of any proposition: Without a sense, or without the thought, a proposition would be an utterly dead and trivial thing. And further it seems clear that no adding of inorganic signs can make the proposition live. And the conclusion which one draws from this is that what must be added to the dead signs in order to make a live proposition is something immaterial, with properties different from all mere signs.

But if we had to name anything which is the life of the sign, we should have to say that it was its *use*.

If the meaning of the sign (roughly, that which is of importance about the sign) is an image built up in our minds when we see or hear the sign, then first let us adopt the method we just described of replacing this mental image by some outward object seen, e.g. a painted or modelled image. Then why should the written sign plus this painted image be alive if the written sign alone was dead?—In fact, as soon as you think of replacing the mental image by, say, a painted one, and as soon as the image thereby loses its occult character, it ceases to seem to impart any life to the sentence at all. (It was in fact just the occult character of the mental process which you needed for your purposes.)

The mistake we are liable to make could be expressed thus: We are looking for the use of a sign, but we look for it as though it were an object *co-existing* with the sign. (One of the reasons for this mistake is again that we are looking for a "thing corresponding to a substantive.")

The sign (the sentence) gets its significance from the system of signs, from the language to which it belongs. Roughly: understanding a sentence means understanding a language.

As a part of the system of language, one may say, the sentence has life. But one is tempted to imagine that which gives the sentence life as something in an occult sphere, accompanying the sentence. But whatever accompanied it would for us just be an-other sign.

It seems at first sight that that which gives to thinking its pe-culiar character is that it is a train of mental states, and it seems that what is queer and difficult to understand about thinking is the processes which happen in the medium of the mind, pro-cesses possible only in this medium. The comparison which forces itself upon us is that of the mental medium with the pro-toplasm of a cell, say, of an amoeba. We observe certain actions of the amoeba, its taking food by extending arms, its splitting up into similar cells, each of which grows and behaves like the origi-nal one. We say "of what a queer nature the protoplasm must be to act in such a way", and perhaps we say that no physical mech-anism could behave in this way, and that the mechanism of the amoeba must be of a totally different kind. In the same way we are tempted to say "the mechanism of the mind must be of a most peculiar kind to be able to do what the mind does". But here we are making two mistakes. For what struck *us* as being queer about thought and thinking was not at all that it had curi-ous effects which we were not yet able to explain (causally). Our problem, in other words, was not a scientific one; but a muddle felt as a problem.

Supposing we tried to construct a mind-model as a result of psychological investigations, a model which, as we should say, would explain the action of the mind. This model would be part of a psychological theory in the way in which a mechanical model of the ether can be part of a theory of electricity. (Such a model, by the way, is always part of the *symbolism* of a theory. Its advantage may be that it can be taken in at a glance and easily held in the mind. It has been said that a model, in a sense, dresses up the pure theory; that the *naked* theory is sentences or equa-tions. This must be examined more closely later on.)

We may find that such a mind-model would have to be very

complicated and intricate in order to explain the observed mental activities; and on this ground we might call the mind a queer kind of medium. But this aspect of the mind does not interest us. The problems which it may set are psychological problems, and the method of their solution is that of natural science.

Now if it is not the causal connections which we are concerned with, then the activities of the mind lie open before us. And when we are worried about the nature of thinking, the puzzlement which we wrongly interpret to be one about the nature of a medium is a puzzlement caused by the mystifying use of our language. This kind of mistake recurs again and again in philosophy; e.g. when we are puzzled about the nature of time, when time seems to us a *queer thing*. We are most strongly tempted to think that here are things hidden, something we can see from the outside but which we can't look into. And yet nothing of the sort is the case. It is not new facts about time which we want to know. All the facts that concern us lie open before us. But it is the use of the substantive "time" which mystifies us. If we look into the grammar of that word, we shall feel that it is no less astounding that man should have conceived of a deity of time than it would be to conceive of a deity of negation or disjunction.

It is misleading then to talk of thinking as of a "mental activity". We may say that thinking is essentially the activity of operating with signs. This activity is performed by the hand, when we think by writing; by the mouth and larynx, when we think by speaking; and if we think by imagining signs or pictures, I can give you no agent that thinks. If then you say that in such cases the mind thinks, I would only draw your attention to the fact that you are using a metaphor, that here the mind is an agent in a different sense from that in which the hand can be said to be the agent in writing.

If again we talk about the locality where thinking takes place we have a right to say that this locality is the paper on which we write or the mouth which speaks. And if we talk of the head or the brain as the locality of thought, this is using the expression "locality of thinking" in a different sense. Let us examine what are the reasons for calling the head the place of thinking. It is not our intention to criticize this form of expression, or to show that it is not appropriate. What we must do is: understand its working, its grammar, e.g. see what relation this grammar has to that of the expression "we think with our mouth", or "we think with a pencil on a piece of paper".

Perhaps the main reason why we are so strongly inclined to talk of the head as the locality of our thoughts is this: the existence of the words "thinking" and "thought" alongside of the words denoting (bodily) activities, such as writing, speaking, etc., makes us look for an activity, different from these but analogous to them, corresponding to the word "thinking". When words in our ordinary language have prima facie analogous grammars we are inclined to try to interpret them analogously; i.e. we try to make the analogy hold throughout.—We say, "The thought is not the same as the sentence; for an English and a French sentence, which are utterly different, can express the same thought". And now, as the sentences are *somewhere*, we look for a place for the thought. (It is as though we looked for the place of the king of which the rules of chess treat, as opposed to the places of the various bits of wood, the kings of the various sets.)—We say, "surely the thought is *something*; it is not nothing"; and all one can answer to this is, that the word "thought" has its *use*, which is of a totally different kind from the use of the word "sentence".

Now does this mean that it is nonsensical to talk of a locality where thought takes place? Certainly not. This phrase has sense if we give it sense. Now if we say "thought takes place in our heads", what is the sense of this phrase soberly understood? I suppose it is that certain physiological processes correspond to our thoughts in such a way that if we know the correspondence we can, by observing these processes, find the thoughts. But in what sense can the physiological processes be said to correspond to thoughts, and in what sense can we be said to get the thoughts from the observation of the brain?

I suppose we imagine the correspondence to have been verified experimentally. Let us imagine such an experiment crudely. It consists in looking at the brain while the subject thinks. And now you may think that the reason why my explanation is going to go wrong is that of course the experimenter gets the thoughts of the subject only *indirectly* by being told them, the subject *expressing* them in some way or other. But I will remove this difficulty by assuming that the subject is at the same time the experimenter, who is looking at his own brain, say by means of a mirror. (The crudity of this description in no way reduces the force of the argument.)

Then I ask you, is the subject-experimenter observing one thing or two things? (Don't say that he is observing one thing both from the inside and from the outside; for this does not remove the difficulty. We will talk of inside and outside later.*) The subject-experimenter is observing a correlation of two phenomena. One of them he, perhaps, calls the *thought*. This may consist of a train of images, organic sensations, or on the other hand of a train of the various visual, tactual and muscular experiences which he has in writing or speaking a sentence.—The other experience is one of seeing his brain work. Both these phenomena could correctly be called "expressions of thought"; and the question "where is the thought itself?" had better, in order to prevent confusion, be rejected as nonsensical. If however we do use the expression "the thought takes place in the head", we have given this expression its meaning by describing the experience which would justify the *hypothesis* that the thought takes places in our heads, by describing the experience which we wish to call "observing thought in our brain".

---

* See pp. 104, 139ff.

We easily forget that the word "locality" is used in many different senses and that there are many different kinds of statements about a thing which in a particular case, in accordance with general usage, we may call specifications of the locality of the thing. Thus it has been said of visual space that its place is in our head; and I think one has been tempted to say this, partly, by a grammatical misunderstanding.

I can say: "in my visual field I see the image of the tree to the right of the image of the tower" or "I see the image of the tree in the middle of the visual field". And now we are inclined to ask "and where do you see the visual field?" Now if the "where" is meant to ask for a locality in the sense in which we have specified the locality of the image of the tree, then I would draw your attention to the fact that you have not yet given this question sense; that is, that you have been proceeding by a grammatical analogy without having worked out the analogy in detail.

In saying that the idea of our visual field being located in our brain arose from a grammatical misunderstanding, I did not mean to say that we could not give sense to such a specification of locality. We could, e.g., easily imagine an experience which we should describe by such a statement. Imagine that we looked at a group of things in this room, and, while we looked, a probe was stuck into our brain and it was found that if the point of the probe reached a particular point in our brain, then a particular small part of our visual field was thereby obliterated. In this way we might co-ordinate points of our brain to points of the visual image, and this might make us say that the visual field was seated in such and such a place in our brain. And if now we asked the question "Where do you see the image of this book?" the answer could be (as above) "To the right of that pencil", or "In the left hand part of my visual field", or again: "Three inches behind my left eye".

But what if someone said "I can assure you I feel the visual image to be two inches behind the bridge of my nose";—what are we to answer him? Should we say that he is not speaking the truth, or that there cannot be such a feeling? What if he asks us "do you know all the feelings there are? How do you know there isn't such a feeling?"

What if the diviner tells us that when he holds the rod he *feels* that the water is five feet under the ground? or that he *feels* that a mixture of copper and gold is five feet under the ground? Suppose that to our doubts he answered: "You can estimate a length when you see it. Why shouldn't I have a different way of estimating it?"

If we understand the idea of such an estimation, we shall get clear about the nature of our doubts about the statements of the diviner, and of the man who said he felt the visual image behind the bridge of his nose.

There is the statement: "this pencil is five inches long", and the statement, "I feel that this pencil is five inches long", and we must get clear about the relation of the grammar of the first statement to the grammar of the second. To the statement "I feel in my hand that the water is three feet under the ground" we should like to answer: "I don't know what this *means*". But the diviner would say: "Surely you know what it means. You know what 'three feet under the ground' means, and you know what 'I feel' means!" But I should answer him: I know what a word means *in certain contexts*. Thus I understand the phrase, "three feet under the ground", say, in the connections "The measurement has shown that the water runs three feet under the ground", "If we dig three feet deep we are going to strike water", "The depth of the water is three feet by the eye". But the use of the expression "a feeling in my hands of water being three feet under the ground" has yet to be explained to me.

We could ask the diviner "how did you learn the meaning of the word 'three feet'? We suppose by being shown such lengths, by having measured them and such like. Were you also taught to talk of a feeling of water being three feet under the ground, a feeling, say, in your hands? For if not, what made you connect the word 'three feet' with a feeling in your hand?" Supposing we had been estimating lengths by the eye, but had never spanned a length. How could we estimate a length in inches by spanning it? I.e., how could we interpret the experience of spanning in inches? The question is: what connection is there between, say, a tactual sensation and the experience of measuring a thing by means of a yard rod? This connection will show us what it means to 'feel that a thing is six inches long'. Supposing the diviner said "I have never learnt to correlate depth of water under the ground with feelings in my hand, but when I have a certain feeling of tension in my hands, the words 'three feet' spring up in my mind." We should answer "This is a perfectly good explanation of what you mean by 'feeling the depth to be three feet', and the statement that you feel this will have neither more, nor less, meaning than your explanation has given it. And if experience shows that the actual depth of the water always agrees with the words 'n feet' which come into your mind, your experience will be very useful for determining the depth of water".—But you see that the meaning of the words "I feel the depth of the water to be n feet" had to be explained; it was not known when the meaning of the words "n feet" in the ordinary sense (i.e. in the ordinary contexts) was known.—We don't say that the man who tells us he feels the visual image two inches behind the bridge of his nose is telling a lie or talking nonsense. But we say that we don't understand the meaning of such a phrase. It combines well-known words, but combines them in a way we don't yet understand. The grammar of this phrase has yet to be explained to us.

The importance of investigating the diviner's answer lies in the fact that we often think we have given a meaning to a statement P if only we assert "I *feel* (or I believe) that P is the case." (We shall talk at a later occasion* of Prof. Hardy saying that

---

* This promise is not kept.—*Edd.*

Goldbach's theorem is a proposition because he can believe that it is true.) We have already said that by merely explaining the meaning of the words "three feet" in the usual way we have not yet explained the sense of the phrase "feeling that water is three feet etc." Now we should not have felt these difficulties had the diviner said that he had *learnt* to estimate the depth of the water, say, by digging for water whenever he had a particular feeling and in this way correlating such feelings with *measurements* of depth. Now we must examine the relation of the process of *learning to estimate* with the act of estimating. The importance of this examination lies in this, that it applies to the relation between learning the meaning of a word and making use of the word. Or, more generally, that it shows the different possible relations between a rule given and its application.

Let us consider the process of estimating a length by the eye: It is extremely important that you should realise that there are a great many different processes which we call "estimating by the eye".

Consider these cases:—

(1) Someone asks "How did you estimate the height of this building?" I answer: "It has four storeys; I suppose each storey is about fifteen feet high; so it must be about sixty feet."

(2) In another case: "I roughly know what a yard at that distance looks like; so it must be about four yards long."

(3) Or again: "I can imagine a tall man reaching to about this point; so it must be about six feet above the ground."

(4) Or: "I don't know; it just looks like a yard."

This last case is likely to puzzle us. If you ask "what happened in this case when the man estimated the length?" the correct answer may be: "he *looked* at the thing and *said* 'it looks one yard long'." This may be all that has happened.

We said before that we should not have been puzzled about the diviner's answer if he had told us that he had *learnt* how to estimate depth. Now learning to estimate may, broadly speaking, be seen in two different relations to the act of estimating: either as a cause of the phenomenon of estimating, or as supplying us with a rule (a table, a chart, or some such thing) which we make use of when we estimate.

Supposing I teach someone the use of the word "yellow" by repeatedly pointing to a yellow patch and pronouncing the word. On another occasion I make him apply what he has learnt by giving him the order, "choose a yellow ball out of this bag". What was it that happened when he obeyed my order? I say "possibly just this: he heard my words and took a yellow ball from the bag". Now you may be inclined to think that this couldn't possibly have been all; and the *kind* of thing that you would suggest is that he imagined something yellow when he *understood* the order, and then chose a ball according to his image. To see that this is not *necessary* remember that I could have given him the order, "Imagine a yellow patch". Would you still be inclined to assume that he first imagines a yellow patch, just *understanding* my order, and then imagines a yellow patch to match the first? (Now I don't say that this is not possible. Only, putting it in this way immediately shows you that it need not happen. This, by the way, illustrates the method of philosophy.)

If we are taught the meaning of the word "yellow" by being given some sort of ostensive definition (a rule of the usage of the word) this teaching can be looked at in two different ways.

A. The teaching is a drill. This drill causes us to associate a yellow image, yellow things, with the word "yellow". Thus when I gave the order "Choose a yellow ball from this bag" the word "yellow" might have brought up a yellow image, or a feeling of recognition when the person's eye fell on the yellow ball. The drill of teaching could in this case be said to have built up a psychical mechanism. This, however, would only be a hypothesis or else a metaphor. We could *compare* teaching with installing an electric connection between a switch and a bulb. The parallel to the connection going wrong or breaking down would then be what we call forgetting the explanation, or the meaning, of the word. (We ought to talk further on about the meaning of "forgetting the meaning of a word"*).

In so far as the teaching brings about the association, feeling of recognition, etc. etc., it is the *cause* of the phenomena of understanding, obeying, etc.; and it is a hypothesis that the process of teaching should be needed in order to bring about these effects. It is conceivable, in this sense, that *all* the processes of understanding, obeying, etc., should have happened without the person ever having been taught the language. (This, just now, seems extremely paradoxical.)

B. The teaching may have supplied us with a rule which is itself involved in the processes of understanding, obeying, etc.; "involved", however, meaning that the expression of this rule forms part of these processes.

We must distinguish between what one might call "a process being *in accordance with* a rule", and, "a process involving a rule" (in the above sense).

Take an example. Some one teaches me to square cardinal numbers; he writes down the row

$$1 \quad 2 \quad 3 \quad 4,$$

and asks me to square them. (I will, in this case again, replace any processes happening 'in the mind' by processes of calculation on the paper.) Suppose, underneath the first row of numbers, I then write:

---

* This he never does.—*Edd.*

1   4   9   16.

What I wrote is in accordance with the general rule of squaring; but it obviously is also in accordance with any number of other rules; and amongst these it is not more in accordance with one than with another. In the sense in which before we talked about a rule being involved in a process, *no* rule was involved in this. Supposing that in order to get to my results I calculated 1×1, 2×2, 3×3, 4×4 (that is, in this case wrote down the calculations); these would again be in accordance with any number of rules. Supposing, on the other hand, in order to get to my results I had written down what you may call "the rule of squaring", say algebraically. In this case this rule was involved in a sense in which no other rule was.

We shall say that the rule is *involved* in the understanding, obeying, etc., if, as I should like to express it, the symbol of the rule forms part of the calculation. (As we are not interested in where the processes of thinking, calculating, take place, we can for our purpose imagine the calculations being done entirely on paper. We are not concerned with the difference: internal, external.)

A characteristic example of the case B would be one in which the teaching supplied us with a table which we actually make use of in understanding, obeying, etc. If we are taught to play chess, we may be taught rules. If then we play chess, these rules need not be involved in the act of playing. But they may be. Imagine, e.g., that the rules were expressed in the form of a table; in one column the shapes of the chessmen are drawn, and in a parallel column we find diagrams showing the 'freedom' (the legitimate moves) of the pieces. Suppose now that the way the game is played involves making the transition from the shape to the possible moves by running one's finger across the table, and then making one of these moves.

Teaching as the hypothetical history of our subsequent actions (understanding, obeying, estimating a length, etc.) drops out of our considerations. The rule which has been taught and is subsequently applied interests us only so far as it is involved in the application. A rule, so far as it interests us, does not act at a distance.

Suppose I pointed to a piece of paper and said to someone: "this colour I call 'red'". Afterwards I give him the order: "now paint me a red patch". I then ask him: "why, in carrying out my order, did you paint just this colour?" His answer could then be: "This colour (pointing to the sample which I have given him) was called red; and the patch I have painted has, as you see, the colour of the sample". He has now given me a reason for carrying out the order in the way he did. Giving a reason for something one did or said means showing a *way* which leads to this action. In some cases it means telling the way which one has gone oneself; in others it means describing a way which leads there and is in accordance with certain accepted rules. Thus when asked, "why did you carry out my order by painting just this colour?" the person could have described the way he had actually taken to arrive at this particular shade of colour. This would have been so if, hearing the word "red", he had taken up the sample I had given him, labelled "red", and had *copied* that sample when painting the patch. On the other hand he might have painted it 'automatically' or from a memory image, but when asked to give the reason he might still point to the sample and show that it matched the patch he had painted. In this latter case the reason given would have been of the second kind; i.e. a justification *post hoc*.

Now if one thinks that there could be no understanding and obeying the order without a previous teaching, one thinks of the teaching as supplying a *reason* for doing what one did; as supplying the road one walks. Now there is the idea that if an order is understood and obeyed there must be a reason for our obeying it as we do; and, in fact, a chain of reasons reaching back to infinity. This is as if one said: "Wherever you are, you must have got there from somewhere else, and to that previous place from another place; and so on *ad infinitum*". (If, on the other hand, you had said, "wherever you are, you *could* have got there from another place ten yards away; and to that other place from a third, ten yards further away, and so on *ad infinitum*", if you had said this you would have stressed the infinite *possibility* of making a step. Thus the idea of an infinite chain of reasons arises out of a confusion similar to this: that a line of a certain length consists of an infinite number of parts because it is indefinitely divisible; i.e., because there is no end to the possibility of dividing it.)

If on the other hand you realize that the chain of *actual* reasons has a beginning, you will no longer be revolted by the idea of a case in which there is *no* reason for the way you obey the order. At this point, however, another confusion sets in, that between reason and cause. One is led into this confusion by the ambiguous use of the word "why". Thus when the chain of reasons has come to an end and still the question "why?" is asked, one is inclined to give a cause instead of a reason. If, e.g., to the question, "why did you paint just this colour when I told you to paint a red patch?" you give the answer: "I have been shown a sample of this colour and the word 'red' was pronounced to me at the same time; and therefore this colour now always comes to my mind when I hear the word 'red'", then you have given a cause for your action and not a reason.

The proposition that your action has such and such a cause is a hypothesis. The hypothesis is well-founded if one has had a number of experiences which, roughly speaking, agree in showing that your action is the regular sequel of certain conditions which we then call causes of the action. In order to know the reason which you had for making a certain statement, for acting in a particular way, etc., no number of agreeing experiences is necessary, and the statement of your reason is not a hypothesis. The difference between the grammars of "reason" and "cause" is quite similar to that between the grammars of "motive" and "cause". Of the cause one can say that one can't *know* it but can only *conjecture* it. On the other hand one often says: "Surely *I* must know why I did it" talking of the *motive*. When I say: "we can only *conjecture* the cause but we *know* the motive" this statement will be seen later on to be a grammatical one. The "can" refers to a *logical* possibility.

The double use of the word "why", asking for the cause and asking for the motive, together with the idea that we can know, and not only conjecture, our motives, gives rise to the confusion that a motive is a cause of which we are immediately aware, a cause 'seen from the inside', or a cause experienced.—Giving a reason is like giving a calculation by which you have arrived at a certain result.

Let us go back to the statement that thinking essentially consists in operating with signs. My point was that it is liable to mislead us if we say 'thinking is a mental activity'. The question what kind of an activity thinking is is analogous to this: "Where does thinking take place?" We can answer: on paper, in our head, in the mind. None of these statements of locality gives *the* locality of thinking. The use of all these specifications is correct, but we must not be misled by the similarity of their linguistic form into a false conception of their grammar. As, e.g., when you say: "Surely, the *real* place of thought is in our head". The same applies to the idea of thinking as an activity. It is correct to say that thinking is an activity of our writing hand, of our larynx, of our head, and of our mind, so long as we understand the grammar of these statements. And it is, furthermore, extremely important to realize how, by misunderstanding the grammar of our expressions, we are led to think of one in particular of these statements as giving the *real* seat of the activity of thinking.

There is an objection to saying that thinking is some such thing as an activity of the hand. Thinking, one wants to say, is part of our 'private experience'. It is not material, but an event in private consciousness. This objection is expressed in the question: "Could a machine think?" I shall talk about this at a later point,*

---

* See p. 144 for a few further remarks on this topic.—*Edd*.

and now only refer you to an analogous question: "Can a machine have toothache?" You will certainly be inclined to say: "A machine can't have toothache". All I will do now is to draw your attention to the use which you have made of the word "can" and to ask you: "Did you mean to say that all our past experience has shown that a machine never had toothache?" The impossibility of which you speak is a logical one. The question is: What is the relation between thinking (or toothache) and the subject which thinks, has toothache, etc.? I shall say no more about this now.

If we say thinking is essentially operating with signs, the first question you might ask is: "What are signs?"—Instead of giving any kind of general answer to this question, I shall propose to you to look closely at particular cases which we should call "operating with signs". Let us look at a simple example of operating with words. I give someone the order: "fetch me six apples from the grocer", and I will describe a way of making use of such an order: The words "six apples" are written on a bit of paper, the paper is handed to the grocer, the grocer compares the word "apple" with labels on different shelves. He finds it to agree with one of the labels, counts from 1 to the number written on the slip of paper, and for every number counted takes a fruit off the shelf and puts it in a bag.—And here you have a case of the use of words. I shall in the future again and again draw your attention to what I shall call language games. These are ways of using signs simpler than those in which we use the signs of our highly complicated everyday language. Language games are the forms of language with which a child begins to make use of words. The study of language games is the study of primitive forms of language or primitive languages. If we want to study the problems of truth and falsehood, of the agreement and disagreement of propositions with reality, of the nature of assertion, assumption, and question, we shall with great advantage look at primitive forms of language in which these forms of thinking appear without the confusing background of highly complicated processes of thought. When we look at such

particular leaves. He was shown different leaves when he learnt the meaning of the word "leaf"; and showing him the particular leaves was only a means to the end of producing 'in him' an idea which we imagine to be some kind of general image. We say that he sees what is in common to all these leaves; and this is true if we mean that he can on being asked tell us certain features or properties which they have in common. But we are inclined to think that the general idea of a leaf is something like a visual image, but one which only contains what is common to all leaves. (Galtonian composite photograph.) This again is connected with the idea that the meaning of a word is an image, or a thing correlated to the word. (This roughly means, we are looking at words as though they all were proper names, and we then confuse the bearer of a name with the meaning of the name.)

(c) Again, the idea we have of what happens when we get hold of the general idea 'leaf', 'plant', etc. etc., is connected with the confusion between a mental state, meaning a state of a hypothetical mental mechanism, and a mental state meaning a state of consciousness (toothache, etc.).

(d) Our craving for generality has another main source: our pre-occupation with the method of science. I mean the method of reducing the explanation of natural phenomena to the smallest possible number of primitive natural laws; and, in mathematics, of unifying the treatment of different topics by using a generalization. Philosophers constantly see the method of science before their eyes, and are irresistibly tempted to ask and answer questions in the way science does. This tendency is the real source of metaphysics, and leads the philosopher into complete darkness. I want to say here that it can never be our job to reduce anything to anything, or to explain anything. Philosophy really *is* 'purely descriptive'.

(Think of such questions as "Are there sense data?" and ask: What method is there of determining this? Introspection?)

Instead of "craving for generality" I could also have said "the contemptuous attitude towards the particular case". If, e.g., someone tries to explain the concept of number and tells us that such and such a definition will not do or is clumsy because it only applies to, say, finite cardinals I should answer that the mere fact that he could have given such a limited definition makes this definition extremely important to us. (Elegance is *not* what we are trying for.) For why should what finite and transfinite numbers have in common be more interesting to us than what distinguishes them? Or rather, I should not have said "why should it be more interesting to us?"—it *isn't*; and this characterizes our way of thinking.

The attitude towards the more general and the more special in logic is connected with the usage of the word "kind" which is liable to cause confusion. We talk of kinds of numbers, kinds of propositions, kinds of proofs; and, also, of kinds of apples, kinds of paper, etc. In one sense what defines the kind are properties, like sweetness, hardness, etc. In the other the different kinds are different grammatical structures. A treatise on pomology may be called incomplete if there exist kinds of apples which it doesn't mention. Here we have a standard of completeness in nature. Supposing on the other hand there was a game resembling that of chess but simpler, no pawns being used in it. Should we call this game incomplete? Or should we call a game more complete than chess if it in some way contained chess but added new elements? The contempt for what seems the less general case in logic springs from the idea that it is incomplete. It is in fact confusing to talk of cardinal arithmetic as something special as opposed to something more general. Cardinal arithmetic bears no mark of incompleteness; nor does an arithmetic which is cardinal and finite. (There are no subtle distinctions between logical forms as there are between the tastes of different kinds of apples.)

If we study the grammar, say, of the words "wishing", "thinking", "understanding", "meaning", we shall not be dissatisfied when we have described various cases of wishing, thinking, etc. If someone said, "surely this is not all that one calls 'wishing'", we should answer, "certainly not, but you can build up more complicated cases if you like." And after all, there is not one definite class of features which characterize all cases of wishing (at least not as the word is commonly used). If on the other hand you wish to give a definition of wishing, i.e., to draw a sharp boundary, then you are free to draw it as you like; and this boundary will never entirely coincide with the actual usage, as this usage has no sharp boundary.

The idea that in order to get clear about the meaning of a general term one had to find the common element in all its applications has shackled philosophical investigation; for it has not only led to no result, but also made the philosopher dismiss as irrelevant the concrete cases, which alone could have helped him to understand the usage of the general term. When Socrates asks the question, "what is knowledge?" he does not even regard it as a *preliminary* answer to enumerate cases of knowledge.* If I wished to find out what sort of thing arithmetic is, I should be very content indeed to have investigated the case of a finite cardinal arithmetic. For

(*a*) this would lead me on to all the more complicated cases,

(*b*) a finite cardinal arithmetic is not incomplete, it has no gaps which are then filled in by the rest of arithmetic.

What happens if from 4 till 4.30 A expects B to come to his room? In one sense in which the phrase "to expect something from 4 to 4.30" is used it certainly does not refer to one process or state

---

* *Theaetetus* 146D–7C.

of mind going on throughout that interval, but to a great many different activities and states of mind. If for instance I expect B to come to tea, what happens *may* be this: At four o'clock I look at my diary and see the name "B" against to-day's date; I prepare tea for two; I think for a moment "does B smoke?" and put out cigarettes; towards 4.30 I begin to feel impatient; I imagine B as he will look when he comes into my room. All this is called "expecting B from 4 to 4.30". And there are endless variations to this process which we all describe by the same expression. If one asks what the different processes of expecting someone to tea have in common, the answer is that there is no single feature in common to all of them, though there are many common features overlapping. These cases of expectation form a family; they have family likenesses which are not clearly defined.

There is a totally different use of the word "expectation" if we use it to mean a particular sensation. This use of the words like "wish", "expectation", etc., readily suggests itself. There is an obvious connection between this use and the one described above. There is no doubt that in many cases if we expect some one, in the first sense, some, or all, of the activities described are accompanied by a peculiar feeling, a tension; and it is natural to use the word "expectation" to mean this experience of tension.

There arises now the question: is this sensation to be called "the sensation of expectation", or "the sensation of expectation that B will come"? If the first case to say that you are in a state of expectation admittedly does not fully describe the situation of expecting that so-and-so will happen. The second case is often rashly suggested as an explanation of the use of the phrase "expecting that so-and-so will happen", and you may even think that with this explanation you are on safe ground, as every further question is dealt with by saying that the sensation of expectation is indefinable.

Now there is no objection to calling a particular sensation "the expectation that B will come". There may even be good practical reasons for using such an expression. Only mark:—if we have explained the meaning of the phrase "expecting that B will come" in this way no phrase which is derived from this by substituting a different name for "B" is thereby explained. One might say that the phrase "expecting that B will come" is not a value of a function "expecting that x will come". To understand this compare our case with that of the function "I eat x". We understand the proposition "I eat a chair" although we weren't specifically taught the meaning of the expression "eating a chair".

The role which in our present case the name "B" plays in the expression "I expect B" can be compared with that which the name "Bright" plays in the expression "Bright's disease"* Compare the grammar of this word, when it denotes a particular kind of disease, with that of the expression "Bright's disease," when it means the disease which Bright has. I will characterize the difference by saying that the word "Bright" in the first case is an index in the complex *name* "Bright's disease"; in the second case I shall call it an argument of the function "x's disease". One may say that an index *alludes* to something, and such an allusion may be justified in all sorts of ways. Thus calling a sensation "the expectation that B will come" is giving it a complex name and "B" possibly alludes to the man whose coming had regularly been preceded by the sensation.

Again we may use the phrase "expectation that B will come" not as a name but as a characteristic of certain sensations. We might, e.g., explain that a certain tension is said to be an expectation that B will come if it is relieved by B's coming. If this is how we use the phrase then it is true to say that we don't know what we expect until our expectation has been fulfilled (cf. Russell). But no one can believe that this is the only way or even the most common way of using the word "expect". If I ask someone "whom do you expect?" and after receiving the answer ask again "Are you sure that you don't expect someone else?" then, in most

---

* See *Tractatus* 5.02.

cases, this question would be regarded as absurd, and the answer will be something like "Surely, I must know whom I expect".

One may characterize the meaning which Russell gives to the word "wishing" by saying that it means to him a kind of hunger.—It is a hypothesis that a particular feeling of hunger will be relieved by eating a particular thing. In Russell's way of using the word "wishing" it makes no sense to say "I wished for an apple but a pear has satisfied me".* But we do sometimes say this, using the word "wishing" in a way different from Russell's. In this sense we can say that the tension of wishing was relieved without the wish being fulfilled; and also that the wish was fulfilled without the tension being relieved. That is, I may, in this sense, become satisfied without my wish having been satisfied.

Now one might be tempted to say that the difference which we are talking about simply comes to this, that in some cases we know what we wish and in others we don't. There are certainly cases in which we say, "I feel a longing, though I don't know what I'm longing for", or, "I feel a fear, but I don't know what I'm afraid of", or again: "I feel fear, but I'm not afraid of anything in particular".

Now we may describe these cases by saying that we have certain sensations not referring to objects. The phrase "not referring to objects" introduces a grammatical distinction. If in characterizing such sensations we use verbs like "fearing", "longing", etc., these verbs will be intransitive; "I fear" will be analogous to "I cry". We may cry about something, but what we cry about is not a constituent of the process of crying; that is to say, we could describe all that happens when we cry without mentioning what we are crying about.

Suppose now that I suggested we should use the expression "I feel fear", and similar ones, in a transitive way only. Whenever before we said "I have a sensation of fear" (intransitively) we will

---

* Cf. Russell, *Analysis of Mind*, III.

now say "I am afraid of something, but I don't know of what". Is there an objection to this terminology?

We may say: "There isn't, except that we are then using the word 'to know' in a queer way". Consider this case:—we have a general undirected feeling of fear. Later on, we have an experience which makes us say, "Now I know what I was afraid of. I was afraid of so-and-so happening". Is it correct to describe my first feeling by an intransitive verb, or should I say that my fear had an object although I did not know that it had one? Both these forms of description can be used. To understand this examine the following example:—It might be found practical to call a certain state of decay in a tooth, not accompanied by what we commonly call toothache, "unconscious toothache" and to use in such a case the expression that we have toothache, but don't know it. It is in just this sense that psychoanalysis talks of unconscious thoughts, acts of volition, etc. Now is it wrong in this sense to say that I have toothache but don't know it? There is nothing wrong about it, as it is just a new terminology and can at any time be retranslated into ordinary language. On the other hand it obviously makes use of the word "to know" in a new way. If you wish to examine how this expression is used it is helpful to ask yourself "what in this case is the process of getting to know like?" "What do we call 'getting to know' or, 'finding out'?"

It isn't wrong, according to our new convention, to say "I have unconscious toothache". For what more can you ask of your notation than that it should distinguish between a bad tooth which doesn't give you toothache and one which does? But the new expression misleads us by calling up pictures and analogies which make it difficult for us to go through with our convention. And it is extremely difficult to discard these pictures unless we are constantly watchful; particularly difficult when, in philosophizing, we contemplate what we *say* about things. Thus, by the expression "unconscious toothache" you may either be misled into thinking that a stupendous discovery has been made, a discovery which in a sense altogether bewilders our understanding; or else you may be extremely puzzled by the expression (the puzzlement of philosophy) and perhaps ask such a question as "How is unconscious toothache possible?" You may then be tempted to deny the possibility of unconscious toothache; but the scientist will tell you that it is a proved fact that there is such a thing, and he will say it like a man who is destroying a common prejudice. He will say: "Surely it's quite simple; there are other things which you don't know of, and there can also be toothache which you don't know of. It is just a new discovery". You won't be satisfied, but you won't know what to answer. This situation constantly arises between the scientist and the philosopher.

In such a case we may clear the matter up by saying: "Let's see how the word 'unconscious', 'to know', etc. etc., is used in *this* case, and how it's used in others". *How far does the analogy between these uses go?* We shall also try to construct new notations, in order to break the spell of those which we are accustomed to.

We said that it was a way of examining the grammar (the use) of the word "to know", to ask ourselves what, in the particular case we are examining, we should call "getting to know". There is a temptation to think that this question is only vaguely relevant, if relevant at all, to the question: "what is the meaning of the word 'to know'?" We seem to be on a side-track when we ask the question "What is it like in this case 'to get to know'?" But this question really is a question concerning the grammar of the word "to know", and this becomes clearer if we put it in the form: "What do we *call* 'getting to know'?" It is part of the grammar of the word "chair" that *this* is what we call "to sit on a chair", and it is part of the grammar of the word "meaning" that *this* is what we call "explanation of a meaning"; in the same way to explain my criterion for another person's having toothache is to give a grammatical explanation about the word "toothache" and, in this sense, an explanation concerning the meaning of the word "toothache".

When we learnt the use of the phrase "so-and-so has toothache" we were pointed out certain kinds of behaviour of those who were said to have toothache. As an instance of these kinds of behaviour let us take holding your cheek. Suppose that by observation I found that in certain cases whenever these first criteria told me a person had toothache, a red patch appeared on the person's cheek. Supposing I now said to someone "I see A has toothache, he's got a red patch on his cheek". He may ask me "How do you know A has toothache when you see a red patch?" I should then point out that certain phenomena had always coincided with the appearance of the red patch.

Now one may go on and ask: "How do you know that he has got toothache when he holds his cheek?" The answer to this might be, "I say, *he* has toothache when he holds his cheek because I hold my cheek when I have toothache". But what if we went on asking:—"And why do you suppose that toothache corresponds to his holding his cheek just because your toothache corresponds to his holding your cheek?" You will be at a loss to answer this question, and find that here we strike rock bottom, that is we have come down to conventions. (If you suggest as an answer to the last question that, whenever we've seen people holding their cheeks and asked them what's the matter, they have answered, "I have toothache",—remember that this experience only co-ordinates holding your cheek with saying certain words.)

Let us introduce two antithetical terms in order to avoid certain elementary confusions: To the question "How do you know that so-and-so is the case?", we sometimes answer by giving '*criteria*' and sometimes by giving '*symptoms*'. If medical science calls angina an inflammation caused by a particular bacillus, and we ask in a particular case "why do you say this man has got angina?" then the answer "I have found the bacillus so-and-so in his blood" gives us the criterion, or what we may call the defining criterion of angina. If on the other hand the answer was, "His throat is inflamed", this might give us a symptom of angina.

I call "symptom" a phenomenon of which experience has taught us that it coincided, in some way or other, with the phenomenon which is our defining criterion. Then to say "A man has angina if this bacillus is found in him" is a tautology or it is a loose way of stating the definition of "angina". But to say, "A man has angina whenever he has an inflamed throat" is to make a hypothesis.

In practice, if you were asked which phenomenon is the defining criterion and which is a symptom, you would in most cases be unable to answer this question except by making an arbitrary decision *ad hoc*. It may be practical to define a word by taking one phenomenon as the defining criterion, but we shall easily be persuaded to define the word by means of what, according to our first use, was a symptom. Doctors will use names of diseases without ever deciding which phenomena are to be taken as criteria and which as symptoms; and this need not be a deplorable lack of clarity. For remember that in general we don't use language according to strict rules—it hasn't been taught us by means of strict rules, either. *We*, in our discussions on the other hand, constantly compare language with a calculus proceeding according to exact rules.

This is a very one-sided way of looking at language. In practice we very rarely use language as such a calculus. For not only do we not think of the rules of usage—of definitions, etc.—while using language, but when we are asked to give such rules, in most cases we aren't able to do so. We are unable clearly to circumscribe the concepts we use; not because we don't know their real definition, but because there is no real 'definition' to them. To suppose that there *must* be would be like supposing that whenever children play with a ball they play a game according to strict rules.

When we talk of language as a symbolism used in an exact calculus, that which is in our mind can be found in the sciences and in mathematics. Our ordinary use of language conforms to this standard of exactness only in rare cases. Why then do we in philosophizing constantly compare our use of words with one following exact rules? The answer is that the puzzles which we try to remove always spring from just this attitude towards language.

Consider as an example the question "What is time?" as Saint Augustine and others have asked it. At first sight what this question asks for is a definition, but then immediately the question arises: "What should we gain by a definition, as it can only lead us to other undefined terms?" And why should one be puzzled just by the lack of a definition of time, and not by the lack of a definition of "chair"? Why shouldn't we be puzzled in all cases where we haven't got a definition? Now a definition often clears up the *grammar* of a word. And in fact it is the grammar of the word "time" which puzzles us. We are only expressing this puzzlement by asking a slightly misleading question, the question: "What is . . . ?" This question is an utterance of unclarity, of mental discomfort, and it is comparable with the question "Why?" as children so often ask it. This too is an expression of a mental discomfort, and doesn't necessarily ask for either a cause or a reason. (Hertz, *Principles of Mechanics*.) Now the puzzlement about the grammar of the word "time" arises from what one might call apparent contradictions in that grammar.

It was such a "contradiction" which puzzled Saint Augustine when he argued: How is it possible that one should measure time? For the past can't be measured, as it is gone by; and the future can't be measured because it has not yet come. And the present can't be measured for it has no extension.

The contradiction which here seems to arise could be called a conflict between two different usages of a word; in this case the word "measure". Augustine, we might say, thinks of the process of measuring a *length*: say, the distance between two marks on a travelling band which passes us, and of which we can only see a tiny bit (the present) in front of us. Solving this puzzle will consist in comparing what we mean by "measurement" (the grammar of the word "measurement") when applied to a distance on a travelling band with the grammar of that word when applied to time. The problem may seem simple, but its extreme difficulty is due to the fascination which the analogy between two similar structures in our language can exert on us. (It is helpful here to remember that it is sometimes almost impossible for a child to believe that one word can have two meanings.)

Now it is clear that this problem about the concept of time asks for an answer given in the form of strict rules. The puzzle is about rules.—Take another example: Socrates' question "What is knowledge?" Here the case is even clearer, as the discussion begins with the pupil giving an example of an exact definition, and then analogous to this a definition of the word "knowledge" is asked for. As the problem is put, it seems that there is something wrong with the ordinary use of the word "knowledge". It appears we don't know what it means, and that therefore, perhaps, we have no right to use it. We should reply: "There is no one exact usage of the word 'knowledge'; but we can make up several such usages, which will more or less agree with the ways the word is actually used".

The man who is philosophically puzzled sees a law in the way a word is used, and, trying to apply this law consistently, comes up against cases where it leads to paradoxical results. Very often the way the discussion of such a puzzle runs is this: First the question is asked "What is time?" This question makes it appear that what we want is a definition. We mistakenly think that a definition is what will remove the trouble (as in certain states of indigestion we feel a kind of hunger which cannot be removed by eating). The question is then answered by a wrong definition; say: "Time is the motion of the celestial bodies". The next step is to see that this definition is unsatisfactory. But this only means that we don't use the word "time" synonymously with "motion of the celestial bodies". However in saying that the first definition is wrong, we are now tempted to think that we must replace it by a different one, the correct one.

Compare with this the case of the definition of number. Here the explanation that a number is the same thing as a numeral satisfies that first craving for a definition. And it is very difficult not to ask: "Well, if it isn't the numeral, *what is* it?"

Philosophy, as we use the word, is a fight against the fascination which forms of expression exert upon us.

I want you to remember that words have those meanings which we have given them; and we give them meanings by explanations. I may have given a definition of a word and used the word accordingly, or those who taught me the use of the word may have given me the explanation. Or else we might, by the explanation of a word, mean the explanation which, on being asked, we are ready to give. That is, if we *are* ready to give any explanation; in most cases we aren't. Many words in this sense then don't have a strict meaning. But this is not a defect. To think it is would be like saying that the light of my reading lamp is no real light at all because it has no sharp boundary.

Philosophers very often talk about investigating, analysing, the meaning of words. But let's not forget that a word hasn't got a meaning given to it, as it were, by a power independent of us, so that there could be a kind of scientific investigation into what the word *really* means. A word has the meaning someone has given to it.

There are words with several clearly defined meanings. It is easy to tabulate these meanings. And there are words of which one might say: They are used in a thousand different ways which gradually merge into one another. No wonder that we can't tabulate strict rules for their use.

It is wrong to say that in philosophy we consider an ideal language as opposed to our ordinary one. For this makes it appear as though we thought we could improve on ordinary language. But ordinary language is all right. Whenever we make up 'ideal languages' it is not in order to replace our ordinary language by them; but just to remove some trouble caused in someone's mind by thinking that he has got hold of the exact use of a common word. That is also why our method is not merely to enumerate actual usages of words, but rather deliberately to invent new ones, some of them because of their absurd appearance.

When we say that by our method we try to counteract the misleading effect of certain analogies, it is important that you should understand that the idea of an analogy being misleading is nothing sharply defined. No sharp boundary can be drawn round the cases in which we should say that a man was misled by an analogy. The use of expressions constructed on analogical patterns stresses analogies between cases often far apart. And by doing this these expressions may be extremely useful. It is, in most cases, impossible to show an exact point where an analogy begins to mislead us. Every particular notation stresses some particular point of view. If, e.g., we call our investigations "philosophy", this title, on the one hand, seems appropriate, on the other hand it certainly has misled people. (One might say that the subject we are dealing with is one of the heirs of the subject which used to be called "philosophy".) The cases in which particularly we wish to say that someone is misled by a form of expression are those in which we would say: "he wouldn't talk as he does if he were aware of this difference in the grammar of such-and-such words, or if he were aware of this other possibility of expression" and so on. Thus we may say of some philosophizing mathematicians that they are obviously not aware of the difference between the many different usages of the word "proof"; and that they are not clear about the difference between the uses of the word "kind", when they talk of kinds of numbers, kinds of proofs, as though the word "kind" here meant the same thing as in the context "kinds of apples". Or, we may say, they are not aware of the different *meanings* of the word "discovery", when in one case we talk of the discovery of the construction of the pentagon and in the other case of the discovery of the South Pole.

Now when we distinguish a transitive and an intransitive use of such words as "longing", "fearing", "expecting", etc., we said that some one might try to smooth over our difficulties by saying: "The difference between the two cases is simply that in one case we know what we are longing for and in the other we don't". Now who says this, I think, obviously doesn't see that the difference which he tried to explain away reappears when we carefully consider the use of the word "to know" in the two cases. The expression "the difference is simply . . ." makes it appear as though we had analysed the case and found a simple analysis; as when we point out that two substances with very different names hardly differ in composition.

We said in this case that we might use both expressions: "we feel a longing" (where "longing" is used intransitively) and "we feel a longing and don't know what we are longing for". It may seem queer to say that we may correctly use either of two forms of expression which seem to contradict each other; but such cases are very frequent.

Let us use the following example to clear this up. We say that the equation $x^2 = -1$ has the solution $\pm \sqrt{-1}$. There was a time when one said that this equation had no solution. Now this statement, whether agreeing or disagreeing with the one which told us the solutions, certainly hasn't its multiplicity. But we can easily give it that multiplicity by saying that an equation $x^2 + ax + b = 0$ hasn't got a solution but comes α near to the nearest solution which is β. Analogously we can say either "A straight line always intersects a circle; sometimes in real, sometimes in complex points", or, "A straight line either intersects a circle, or it doesn't and is α far from doing so". These two statements mean exactly the same. They will be more or less satisfactory according to the way a man wishes to look at it. He may wish to make the difference between intersecting and not intersecting as inconspicuous as possible. Or on the other hand he may wish to stress it; and either tendency may be justified, say, by his particular practical purposes. But this may not be the reason at all why he prefers one form of expression to the other. Which form he prefers, and whether he has a preference at all, often depends on general, deeply rooted, tendencies of his thinking.

(Should we say that there are cases when a man despises another man and doesn't know it; or should we describe such cases by saying that he doesn't despise him but unintentionally behaves towards him in a way—speaks to him in a tone of voice, etc.—which in general would go together with despising him? Either form of expression is correct; but they may betray different tendencies of the mind.)

Let us revert to examining the grammar of the expressions "to wish", "to expect", "to long for", etc., and consider that most important case in which the expression "I wish so and so to happen" is the direct description of a conscious process. That is to say, the case in which we should be inclined to answer the question "Are you sure that it is this you wish?" by saying: "Surely I must know what I wish". Now compare this answer to the one which most of us would give to the question: "Do you know the ABC?" Has the emphatic assertion that you know it a sense analogous to that of the former assertion? Both assertions in a way brush aside the question. But the former doesn't wish to say "Surely I know such a simple thing as this" but rather: "The question which you asked me makes no sense". We might say: We adopt in this case a wrong method of brushing aside the question. "Of course I know" could here be replaced by "Of course, there is no doubt" and this interpreted to mean "It makes, in this case, no sense of talk of a doubt". In this way the answer "Of course I know what I wish" can be interpreted to be a grammatical statement.

It is similar when we ask, "Has this room a length?", and someone answers: "Of course it has". He might have answered, "Don't ask nonsense". On the other hand "The room has length" can be used as a grammatical statement. It then says that a sentence of the form "The room is ——— feet long" makes sense.

A great many philosophical difficulties are connected with

that sense of the expressions "to wish", "to think", etc., which we are now considering. These can all be summed up in the question: "How can one think what is not the case?"

This is a beautiful example of a philosophical question. It asks "How can one . . . ?" and while this puzzles us we must admit that nothing is easier than to think what is not the case. I mean, this shows us again that the difficulty which we are in does not arise through our inability to imagine how thinking something is done; just as the philosophical difficulty about the measurement of time did not arise through our inability to imagine how time was actually measured. I say this because sometimes it almost seems as though our difficulty were one of remembering exactly what happened when we thought something, a difficulty of introspection, or something of the sort; whereas in fact it arises when we look at the facts through the medium of a misleading form of expression.

"How can one think what is not the case? If I think that King's College is on fire when it is not on fire, the fact of its being on fire does not exist. Then how can I think it? How can we hang a thief who doesn't exist?" Our answer could be put in this form: "I can't hang him when he doesn't exist; but I can look for him when he doesn't exist".

We are here misled by the substantives "object of thought" and "fact", and by the different meanings of the word "exist".

Talking of the fact as a "complex of objects" springs from this confusion (cf. *Tractatus Logico-Philosophicus*). Supposing we asked: "How can one imagine what does not exist?" The answer seems to be: "If we do, we imagine non-existent combinations of existing elements". A centaur doesn't exist, but a man's head and torso and arms and a horse's legs do exist. "But can't we imagine an object utterly different from any one which exists?"—We should be inclined to answer: "No; the elements, individuals, must exist. If redness, roundness and sweetness did not exist, we could not imagine them".

But what do you mean by "redness exists"? My watch exists, if it hasn't been pulled to pieces, if it hasn't been *destroyed*. What would we call "destroying redness"? We might of course mean destroying all red objects; but would this make it impossible to imagine a red object? Supposing to this one answered: "But surely, red objects must have existed and you must have seen them if you are able to imagine them"?—But how do you know that this is so? Suppose I said "Exerting a pressure on your eye-ball produces a red image". Couldn't the way by which you first became acquainted with red have been this? And why shouldn't it have been just imagining a red patch? (The difficulty which you may feel here will have to be discussed at a later occasion.*

We may now be inclined to say: "As the fact which would make our thought true if it existed does not always exist, it is not the *fact* which we think". But this just depends upon how I wish to use the word "fact". Why shouldn't I say: "I believe the fact that the college is on fire"? It is just a clumsy expression for saying: "I believe that the college is on fire". To say "It is not the fact which we believe", is itself the result of a confusion. We think we are saying something like: "It isn't the sugar-cane which we eat but the sugar", "It isn't Mr. Smith who hangs in the gallery, but his picture".

The next step we are inclined to take is to think that as the object of our thought isn't the fact it is a shadow of the fact. There are different names for this shadow, e.g. "proposition", "sense of the sentence".

But this doesn't remove our difficulty. For the question now is: "How can something be the shadow of a fact which doesn't exist?"

I can express our trouble in a different form by saying: "How can we know what the shadow is a shadow of?"—The shadow would be some sort of portrait; and therefore I can restate our

---

* He does not do this.—*Edd.*

problem by asking: "What makes a portrait a portrait of Mr. N?" The answer which might first suggest itself is: "The similarity between the portrait and Mr. N". This answer in fact shows what we had in mind when we talked of the shadow of a fact. It is quite clear, however, that similarity does not constitute our idea of a portrait; for it is in the essence of this idea that it should make sense to talk of a good or a bad portrait. In other words, it is essential that the shadow should be capable of representing things as in fact they are not.

An obvious, and correct, answer to the question "What makes a portrait the portrait of so-and-so?" is that it is the *intention*. But if we wish to know what it means "intending this to be a portrait of so-and-so" let's see what actually happens when we intend this. Remember the occasion when we talked of what happened when we expect some one from four to four-thirty. To intend a picture to be the portrait of so-and-so (on the part of the painter, e.g.) is neither a particular state of mind nor a particular mental process. But there are a great many combinations of actions and states of mind which we should call "intending . . ." It might have been that he was told to paint a portrait of N, and sat down before N, going through certain actions which we call "copying N's face". One might object to this by saying that the essence of copying is the intention to copy. I should answer that there are a great many different processes which we call "copying something". Take an instance. I draw an ellipse on a sheet of paper and ask you to copy it. What characterizes the process of copying? For it is clear that it isn't the fact that you draw a similar ellipse. You might have tried to copy it and not succeeded; or you might have drawn an ellipse with a totally different intention, and it happened to be like the one you should have copied. So what do you do when you try to copy the ellipse? Well, you look at it, draw something on a piece of paper, perhaps measure what you have drawn, perhaps you curse if you find it doesn't agree with the model; or perhaps you say "I am going to copy this ellipse" and just draw an ellipse like it. There are an endless variety of actions and words, having a family likeness to each other, which we call "trying to copy".

Suppose we said "that a picture is a portrait of a particular object consists in its being derived from that object in a particular way". Now it is easy to describe what we should call processes of deriving a picture from an object (roughly speaking, processes of projection). But there is a peculiar difficulty about admitting that any such process is what we call "intentional representation". For describe whatever process (activity) of projection we may, there is a way of reinterpreting this projection. Therefore—one is tempted to say—such a process can never be the intention itself. For we could always have intended the opposite by reinterpreting the process of projection. Imagine this case: We give someone an order to walk in a certain direction by pointing or by drawing an arrow which points in the direction. Suppose drawing arrows is the language in which generally we give such an order. Couldn't such an order be interpreted to mean that the man who gets it is to walk in the direction opposite to that of the arrow? This could obviously be done by adding to our arrow some symbols which we might call "an *interpretation*". It is easy to imagine a case in which, say to deceive someone, we might make an arrangement that an order should be carried out in the sense opposite to its normal one. The symbol which adds the interpretation to our original arrow could, for instance, be another arrow. Whenever we interpret a symbol in one way or another, the interpretation is a new symbol added to the old one.

Now we might say that whenever we give someone an order by showing him an arrow, and don't do it 'mechanically' (without thinking), we *mean* the arrow in one way or another. And this process of meaning, of whatever kind it may be, can be represented by another arrow (pointing in the same or the opposite sense to the first). In this picture which we make of 'meaning and

saying' it is essential that we should imagine the processes of saying and meaning to take place in two different spheres.

Is it then correct to say that no arrow could be the meaning,
as every arrow could be meant the opposite way?—Suppose we
write down the scheme of saying and meaning by a column of
arrows one below the other.

Then if this scheme is to serve our purpose at all, it must show us
which of the three levels is the level of meaning. I can, e.g., make
a scheme with three levels, the bottom level always being the
level of meaning. But adopt whatever model or scheme you may,
it will have a bottom level, and there will be no such thing as an
interpretation of that. To say in this case that every arrow can
still be interpreted would only mean that I *could* always make a
different model of saying and meaning which had one more level
than the one I am using.

Let us put it in this way:—What one wishes to say is: "Every
sign is capable of interpretation; but the *meaning* mustn't be capable of interpretation. It is the last interpretation." Now I assume that you take the meaning to be a process accompanying
the saying, and that it is translatable into, and so far equivalent
to, a further sign. You have therefore further to tell me what you
take to be the distinguishing mark between *a sign* and *the meaning.* If you do so, e.g., by saying that the meaning is the arrow
which you *imagine* as opposed to any which you may draw or
produce in any other way, you thereby say that you will call no
further arrow an interpretation of the one which you have imagined.

All this will become clearer if we consider what it is that really
happens when we say a thing and mean what we say.—Let us ask
ourselves: If we say to someone "I should be delighted to see you"

and mean it, does a conscious process run alongside these words, a process which could itself be translated into spoken words? This will hardly ever be the case.

But let us imagine an instance in which it does happen. Supposing I had a habit of accompanying every English sentence which I said aloud by a German sentence spoken to myself inwardly. If then, for some reason or other, you call the silent sentence the meaning of the one spoken aloud, the process of meaning accompanying the process of saying would be one which could itself be translated into outward signs. Or, *before* any sentence which we say aloud we say its meaning (whatever it may be) to ourselves in a kind of aside. An example at least similar to the case we want would be saying one thing and at the same time seeing a picture before our mind's eye which is the meaning and agrees or disagrees with what we say. Such cases and similar ones exist, but they are not at all what happens as a rule when we say something and mean it, or mean something else. There are, of course, real cases in which what we call meaning is a definite conscious process accompanying, preceding, or following the verbal expression and itself a verbal expression of some sort or translatable into one. A typical example of this is the 'aside' on the stage.

But what tempts us to think of the meaning of what we say as a process essentially of the kind which we have described is the analogy between the forms of expression:

"to say something"
"to mean something".

which seem to refer to two parallel processes.

A process accompanying our words which one might call the "process of meaning them" is the modulation of the voice in which we speak the words; or one of the processes similar to this, like the play of facial expression. These accompany the spoken words not in the way a German sentence might accompany an English sentence, or writing a sentence to accompany speaking a sentence; but in the sense in which the tune of a song accompanies its words. This tune corresponds to the 'feeling' with which we say the sentence. And I wish to point out that this feeling is the expression with which the sentence is said, or something similar to this expression.

Let us revert to our question: "What is the object of a thought?" (e.g. when we say, "I think that King's College is on fire").

The question as we put it is already the expression of several confusions. This is shown by the mere fact that it almost sounds like a question of physics; like asking: "What are the ultimate constituents of matter?" (It is a typically metaphysical question; the characteristic of a metaphysical question being that we express an unclarity about the grammar of words in the *form* of a scientific question.)

One of the origins of our question is the two-fold use of the propositional function "I think x". We say, "I think that so-and-so will happen" or "that so-and-so is the case", and also "I think just the same *thing* as he"; and we say "I expect him", and also "I expect that he will come". Compare "I expect him" and "I shoot him". We can't shoot him if he isn't there. This is how the question arises: "How can we expect something that is not the case?", "How can we expect a fact which does not exist?"

The way out of this difficulty seems to be: what we expect is not the fact, but a shadow of the fact; as it were, the next thing to the fact. We have said that this is only pushing the question one step further back. There are several origins to this idea of a shadow. One of them is this: we say "Surely two sentences of different languages can have the same sense"; and we argue, "therefore the sense is not the same as the sentence", and ask the question "What is the sense?" And we make of 'it' a shadowy being, one of the many which we create when we wish to give meaning to substantives to which no material objects correspond.

Another source of the idea of a shadow being the object of our thought is this: We imagine the shadow to be a picture the intention of which *cannot be questioned*, that is, a picture which we don't interpret in order to understand it, but which we understand without interpreting it. Now there are pictures of which we should say that we interpret them, that is, translate them into a different kind of picture, in order to understand them; and pictures of which we should say that we understand them immediately, without any further interpretation. If you see a telegram written in cipher, and you know the key to this cipher, you will, in general, not say that you understand the telegram before you have translated it into ordinary language. Of course you have only replaced one kind of symbols by another; and yet if now you read the telegram in your language no further process of interpretation will take place.—Or rather, you may now, in certain cases, again translate this telegram, say into a picture; but then too you have only replaced one set of symbols by another.

The shadow, as we think of it, is some sort of a picture; in fact, something very much like an image which comes before our mind's eye; and this again is something not unlike a painted representation in the ordinary sense. A source of the idea of the shadow certainly is the fact that in some cases saying, hearing, or reading a sentence brings images before our mind's eye, images which more or less strictly correspond to the sentence, and which are therefore, in a sense, translations of this sentence into a pictorial language.—But it is absolutely essential for the picture which we imagine the shadow to be that it is what I shall call a "picture by similarity". I don't mean by this that it is a picture similar to what it is intended to represent, but that it is a picture which is correct only when it is similar to what it represents. One might use for this kind of picture the word "copy". Roughly speaking, copies are good pictures when they can easily be mistaken for what they represent.

A plane projection of one hemisphere of our terrestrial globe is not a picture by similarity or a copy in this sense. It would be conceivable that I portrayed some one's face by projecting it in some queer way, though correctly according to the adopted rule of projection, on a piece of paper, in such a way that no one would normally call the projection "a good portrait of so-and-so" because it would not look a bit like him.

If we keep in mind the possibility of a picture which, though correct, has no similarity with its object, the interpolation of a shadow between the sentence and reality loses all point. For now the sentence itself can serve as such a shadow. The sentence is just such a picture, which hasn't the slightest similarity with what it represents. If we were doubtful about how the sentence "King's College is on fire" can be a picture of King's College on fire, we need only ask ourselves: "How should we explain what the sentence means?" Such an explanation might consist of ostensive definitions. We should say, e.g., "this is King's College" (pointing to the building), "this is a fire" (pointing to a fire). This shews you the way in which words and things may be connected.

The idea that that which we wish to happen must be present as a shadow in our wish is deeply rooted in our forms of expression. But, in fact, we might say that it is only the next best absurdity to the one which we should really like to say. If it weren't too absurd we should say that the fact which we wish for must be present in our wish. For how can we wish *just this* to happen if just this isn't present in our wish? It is quite true to say: The mere shadow won't do; for it stops short before the object; and we want the wish to contain the object itself.—We want that the wish that Mr. Smith should come into this room should wish that just *Mr. Smith*, and no substitute, should do the *coming*, and no substitute for that, *into my room*, and no substitute for that. But this is exactly what we said.

Our confusion could be described in this way: Quite in accordance with our usual form of expression we think of the fact which we wish for as a thing which is not yet here, and to which, therefore, we cannot point. Now in order to understand the grammar of our expression "object of our wish" let's just consider the answer which we give to the question: "What is the object of your wish?" The answer to this question of course is "I wish that so-and-so should happen". Now what would the answer be if we went on asking: "And what is the object of this wish?" It could only consist in a repetition of our previous expression of the wish, or else in a translation into some other form of expression. We might, e.g., state what we wished in other words or illustrate it by a picture, etc., etc. Now when we are under the impression that what we call the object of our wish is, as it were, a man who has not yet entered our room, and therefore can't yet be seen, we imagine that any explanation of what it is we wish is only the next best thing to the explanation which would show *the actual fact*—which, we are afraid, can't yet be shown as it has not yet entered.—It is as though I said to some one "I am expecting Mr. Smith", and he asked me "Who is Mr. Smith?", and I answered, "I can't show him to you now, as he isn't there. All I can show you is a picture of him". It then seems as though I could never entirely explain what I wished until it had actually happened. But of course this is a delusion. The truth is that I needn't be able to give a better explanation of what I wished after the wish was fulfilled than before; for I might perfectly well have shown Mr. Smith to my friend, and have shown him what "coming in" means, and have shown him what my room is, before Mr. Smith came into my room.

Our difficulty could be put this way: We think about things,—but how do these things enter into our thoughts? We think about Mr. Smith; but Mr. Smith need not be present. A picture of him won't do; for how are we to know whom it represents? In fact no substitute for him will do. Then how can he himself be an object of our thoughts? (I am here using the expression "object of our thought" in a way different from that in which I have used it before. I mean now a thing I am thinking *about*, not 'that which I am thinking'.)

We said the connection between our thinking, or speaking, about a man and the man himself was made when, in order to explain the meaning of the word "Mr. Smith" we pointed to him, saying "this is Mr. Smith". And there is nothing mysterious about this connection. I mean, there is no queer mental act which somehow conjures up Mr. Smith in our minds when he really isn't here. What makes it difficult to see that this is the connection is a peculiar form of expression of ordinary language, which makes it appear that the connection between our thought (or the expression of our thought) and the thing we think about must have subsisted *during* the act of thinking.

"Isn't it queer that in Europe we should be able to mean someone who is in America?"—If someone had said "Napoleon was crowned in 1804", and we asked him "Did you mean the man who won the battle of Austerlitz?" he might say "Yes, I meant him". And the use of the past tense "meant" might make it appear as though the idea of Napoleon having won the battle of Austerlitz must have been present in the man's mind when he said that Napoleon was crowned in 1804.

Someone says, "Mr. N. will come to see me this afternoon"; I ask "Do you mean him?" pointing to someone present, and he answers "Yes". In this conversation a connection was established between the word "Mr. N." and Mr. N. But we are tempted to think that while my friend said, "Mr. N. will come to see me", and meant what he said, his mind must have made the connection.

This is partly what makes us think of meaning or thinking as a peculiar *mental activity*; the word "mental" indicating that we mustn't expect to understand how these things work.

What we said of thinking can also be applied to imagining. Someone says, he imagines King's College on fire. We ask him: "How do you know that it's *King's College* you imagine on fire? Couldn't it be a different building, very much like it? In fact, is your imagination so absolutely exact that there might not be a

dozen buildings whose representation your image could be?"—
And still you say: "There's no doubt I imagine King's College
and no other building". But can't saying this be making the very
connection we want? For saying it is like writing the words "Portrait of Mr. So-and-so" under a picture. It might have been that
while you imagined King's College on fire you said the words
"King's College is on fire". But in very many cases you certainly
don't speak explanatory words in your mind while you have the
image. And consider, even if you do, you are not going the whole
way from your image to King's College, but only to the words
"King's College". The connection between these words and
King's College was, perhaps, made at another time.

The fault which in all our reasoning about these matters we
are inclined to make is to think that images and experiences of
all sorts, which are in some sense closely connected with each
other, must be present in our mind at the same time. If we sing a
tune we know by heart, or say the alphabet, the notes or letters
seem to hang together, and each seems to draw the next after it,
as though they were a string of pearls in a box, and by pulling out
one pearl I pulled out the one following it.

Now there is no doubt that, having the visual image of a
string of beads being pulled out of a box through a hole in the
lid, we should be inclined to say: "These beads must all have been
together in the box before". But it is easy to see that this is making a hypothesis. I should have had the same image if the beads
had gradually come into existence in the hole of the lid. We easily overlook the distinction between stating a conscious mental
event, and making a hypothesis about what one might call the
mechanism of the mind. All the more as such hypotheses or pictures of the working of our mind are embodied in many of the
forms of expression of our everyday language. The past tense
"meant" in the sentence "I meant the man who won the battle of
Austerlitz" is part of such a picture, the mind being conceived as

a place in which what we remember is kept, stored, before we express it. If I whistle a tune I know well and am interrupted in the middle, if then someone asks me "did you know how to go on?" I should answer "yes, I did". What sort of process is this *knowing how to go on*? It might appear as though the whole continuation of the tune had to be present while I knew how to go on.

Ask yourself such a question as: "How long does it take to know how to go on?" Or is it an instantaneous process? Aren't we making a mistake like mixing up the existence of a gramophone record of a tune with the existence of the tune? And aren't we assuming that whenever a tune passes through existence there must be some sort of a gramophone record of it from which it is played?

Consider the following example: A gun is fired in my presence and I say: "This crash wasn't as loud as I had expected". Someone asks me: "How is this possible? Was there a crash, louder than that of a gun, in your imagination?" I must confess that there was nothing of the sort. Now he says: "Then you didn't really expect a louder crash—but perhaps the shadow of one.—And how did you know that it was the shadow of a louder crash?"—Let's see what, in such a case, might really have happened. Perhaps in waiting for the report I opened my mouth, held on to something to steady myself, and perhaps I said: "This is going to be terrible". Then, when the explosion was over: "It wasn't so loud after all".—Certain tensions in my body relax. But what is the connection between these tensions, opening my mouth, etc., and a real louder crash? Perhaps this connection was made by having heard such a crash and having had the experiences mentioned.

Examine expressions like "having an idea in one's mind", "analysing the idea before one's mind". In order not to be misled by them see what really happens when, say, in writing a letter you are looking for the words which correctly express the idea which is "before your mind". To say that we are trying to express the idea which is before our mind is to use a metaphor, one which very naturally suggests itself; and which is all right so long as it doesn't mislead us when we are philosophizing. For when we recall what really happens in such cases we find a great variety of processes more or less akin to each other.—We might be inclined to say that in all such cases, at any rate, we are *guided* by something before our mind. But then the words "guided" and "thing before our mind" are used in as many senses as the words "idea" and "expression of an idea".

The phrase "to express an idea which is before our mind" suggests that what we are trying to express in words is already expressed, only in a different language; that this expression is before our mind's eye; and that what we do is to translate from the mental into the verbal language. In most cases which we call "expressing an idea, etc." something very different happens. Imagine what it is that happens in cases such as this: I am groping for a word. Several words are suggested and I reject them. Finally one is proposed and I say: "That is what I meant!"

(We should be inclined to say that the proof of the impossibility of trisecting the angle with ruler and compasses analyses our idea of the trisection of an angle. But the proof gives us a new idea of trisection, one which we didn't have before the proof constructed it. The proof led us a road *which we were inclined to go*; but it led us away from where we were, and didn't just show us clearly the place where we had been all the time.)

Let us now revert to the point where we said that we gained nothing by assuming that a shadow must intervene between the expression of our thought and the reality with which our thought is concerned. We said that if we wanted a picture of reality the sentence itself *is* such a picture (though not a picture by similarity).

I have been trying in all this to remove the temptation to think that there '*must* be' what is called a mental process of thinking, hoping, wishing, believing, etc., independent of the process of expressing a thought, a hope, a wish, etc. And I want to give you the following rule of thumb: If you are puzzled about the nature of thought, belief, knowledge, and the like, substitute for the thought the expression of the thought, etc. The difficulty which lies in this substitution, and at the same time the whole point of it, is this: the expression of belief, thought, etc., is just a sentence;—and the sentence has sense only as a member of a system of language; as one expression within a calculus. Now we are tempted to imagine this calculus, as it were, as a permanent background to every sentence which we say, and to think that, although the sentence as written on a piece of paper or spoken stands isolated, in the mental act of thinking the calculus is there—all in a lump. The mental act seems to perform in a miraculous way what could not be performed by any act of manipulating symbols. Now when the temptation to think that in some sense the whole calculus must be present at the same time vanishes, there is no more point in *postulating* the existence of a peculiar kind of mental act alongside of our expression. This, of course, doesn't mean that we have shown that peculiar acts of consciousness do not accompany the expressions of our thoughts! Only we no longer say that they *must* accompany them.

"But the expression of our thoughts can always lie, for we may say one thing and mean another". Imagine the many different things which happen when we say one thing and mean another!—Make the following experiment: say the sentence "It is hot in this room", and mean: "it is cold". Observe closely what you are doing.

We could easily imagine beings who do their private thinking by means of 'asides' and who manage their lies by saying one thing aloud, following it up by an aside which says the opposite.

"But meaning, thinking, etc., are private experiences. They are not activities like writing, speaking, etc."—But why shouldn't they be the specific private experiences of writing—the muscular, visual, tactile sensations of writing or speaking?

Make the following experiment: say and mean a sentence, e.g.: "It will probably rain tomorrow". Now think the same thought again, mean what you just meant, but without saying anything (either aloud or to yourself). If thinking that it will rain tomorrow accompanied saying that it will rain tomorrow, then just do the first activity and leave out the second.—If thinking and speaking stood in the relation of the words and the melody of a song, we could leave out the speaking and do the thinking just as we can sing the tune without the words.

But can't one at any rate speak and leave out the thinking? Certainly—but observe what sort of thing you are doing if you speak without thinking. Observe first of all that the process which we might call "speaking and meaning what you speak" is not necessarily distinguished from that of speaking thoughtlessly by what happens *at the time when you speak*. What distinguishes the two may very well be what happens before or after you speak.

Suppose I tried, deliberately, to speak without thinking;—what in fact would I do? I might read out a sentence from a book, trying to read it automatically, that is, trying to prevent myself from following the sentence with images and sensations which otherwise it would produce. A way of doing this would be to concentrate my attention on something else while I was speaking the sentence, e.g., by pinching my skin hard while I was speaking.— Put it this way: Speaking a sentence without thinking consists in switching on speech and switching off certain accompaniments of speech. Now ask yourself: Does thinking the sentence without speaking it consist in turning over the switch (switching on what we previously switched off and vice versa); that is: does thinking the sentence without speaking it now simply consist in keeping on what accompanied the words but leaving out the words? Try to think the thoughts of a sentence without the sentence and see whether this is what happens.

Let us sum up: If we scrutinize the usages which we make of such words as "thinking", "meaning", "wishing", etc., going through this process rids us of the temptation to look for a peculiar act of thinking, independent of the act of expressing our thoughts, and stowed away in some peculiar medium. We are no longer prevented by the established forms of expression from recognizing that the experience of thinking *may* be just the experience of saying, or may consist of this experience plus others which accompany it. (It is useful also to examine the following case: Suppose a multiplication is part of a sentence; ask yourself what it is like to say the multiplication $7 \times 5 = 35$, thinking it, and, on the other hand, saying it without thinking.) The scrutiny of the grammar of a word weakens the position of certain fixed standards of our expression which had prevented us from seeing facts with unbiassed eyes. Our investigation tried to remove this bias, which forces us to think that the facts *must* conform to certain pictures embedded in our language.

"Meaning" is one of the words of which one may say that they have odd jobs in our language. It is these words which cause most philosophical troubles. Imagine some institution: most of its members have certain regular functions, functions which can easily be described, say, in the statutes of the institution. There are, on the other hand, some members who are employed for odd jobs, which nevertheless may be extremely important.—What causes most trouble in philosophy is that we are tempted to describe the use of important 'odd-job' words as though they were words with regular functions.

The reason I postponed talking about personal experience was that thinking about this topic raises a host of philosophical difficulties which threaten to break up all our commonsense notions about what we should commonly call the objects of our experience. And if we were struck by these problems it might seem to us that all we have said about signs and about the various objects we mentioned in our examples may have to go into the melting-pot.

The situation in a way is typical in the study of philosophy; and one sometimes has described it by saying that no philosophical problem can be solved until all philosophical problems are solved; which means that as long as they aren't all solved every new difficulty renders all our previous results questionable. To this statement we can only give a rough answer if we are to speak about philosophy in such general terms. It is, that every new problem which arises may put in question the *position* which our previous partial results are to occupy in the final picture. One then speaks of having to reinterpret these previous results; and we should say: they have to be placed in a different surrounding.

Imagine we had to arrange the books of a library. When we begin the books lie higgledy-piggledy on the floor. Now there would be many ways of sorting them and putting them in their places. One would be to take the books one by one and put each on the shelf in its right place. On the other hand we might take up several books from the floor and put them in a row on a shelf, merely in order to indicate that these books ought to go together in this order. In the course of arranging the library this whole row of books will have to change its place. But it would be wrong to say that therefore putting them together on a shelf was no step towards the final result. In this case, in fact, it is pretty obvious that having put together books which belong together was a definite achievement, even though the whole row of them had to be shifted. But some of the greatest achievements in philosophy could only be compared with taking up some books which seemed to belong together, and putting them on different shelves; nothing more being final about their positions than that they no longer lie side by side. The onlooker who doesn't know the difficulty of the task might well think in such a case that nothing at all had been achieved.—The difficulty in philosophy is to say no more than we know. E.g., to see that when we have put two books together in their right order we have not thereby put them in their final places.

When we think about the relation of the objects surrounding us to our personal experiences of them, we are sometimes tempted to say that these personal experiences are the material of which reality consists. How this temptation arises will become clearer later on.

When we think in this way we seem to lose our firm hold on the objects surrounding us. And instead we are left with a lot of separate personal experiences of different individuals. These personal experiences again seem vague and seem to be in constant flux. Our language seems not to have been made to describe them. We are tempted to think that in order to clear up such matters philosophically our ordinary language is too coarse, that we need a more subtle one.

We seem to have made a discovery—which I could describe by saying that the ground on which we stood and which appeared to be firm and reliable was found to be boggy and unsafe.—That is, this happens when we philosophize; for as soon as we revert to the standpoint of common sense this *general* uncertainty disappears.

This queer situation can be cleared up somewhat by looking at an example; in fact a kind of parable illustrating the difficulty we are in, and also showing the way out of this sort of difficulty: We have been told by popular scientists that the floor on which we stand is not solid, as it appears to common sense, as it has been discovered that the wood consists of particles filling space so thinly that it can almost be called empty. This is liable to perplex us, for in a way of course we know that the floor is solid, or that, if it isn't solid, this may be due to the wood being rotten but not to its being composed of electrons. To say, on this latter ground, that the floor is not solid is to misuse language. For even if the particles were as big as grains of sand, and as close together as these are in a sandheap, the floor would not be solid if it were composed of them in the sense in which a sandheap is composed of grains. Our perplexity was based on a misunderstanding; the picture of the thinly filled space had been wrongly *applied*. For this picture of the structure of matter was meant to explain the very phenomenon of solidity.

As in this example the word "solidity" was used wrongly and it seemed that we had shown that nothing really was solid, just in this way, in stating our puzzles about the *general vagueness* of sense-experience, and about the flux of all phenomena, we are using the words "flux" and "vagueness" wrongly, in a typically metaphysical way, namely without an antithesis; whereas in their correct and everyday use vagueness is opposed to clearness, flux to stability, inaccuracy to accuracy, and *problem* to *solution*. The very word "problem", one might say, is misapplied when used for our philosophical troubles. These difficulties, as long as they are seen as problems, are tantalizing, and appear insoluble.

There is a temptation for me to say that only my own experience is real: "I know that *I* see, hear, feel pains, etc., but not that anyone else does: I can't know this, because I am I and they are they."

On the other hand I feel ashamed to say to anyone that my experience is the only real one; and I know that he will reply that he could say exactly the same thing about his experience. This seems to lead to a silly quibble. Also I am told: "If you pity someone for having pains, surely you must at least *believe* that he has pains". But how can I even *believe* this? How can these words make sense to me? How could I even have come by the idea of another's experience if there is no possibility of any evidence for it?

But wasn't this a queer question to ask? *Can't* I believe that someone else has pains? Is it not quite easy to believe this?—Is it an answer to say that things are as they appear to common sense?—Again, needless to say, we don't feel these difficulties in ordinary life. Nor is it true to say that we feel them when we scrutinize our experiences by introspection, or make scientific investigations about them. But somehow, when we look at them in a certain way, our expression is liable to get into a tangle. It seems to us as though we had either the wrong pieces, or not enough of them, to put together our jig-saw puzzle. But they are all there, only all mixed up; and there is a further analogy between the jig-saw puzzle and our case: It's no use trying to apply force in fitting pieces together. All we should do is to look at them *carefully* and arrange them.

There are propositions of which we may say that they describe facts in the material world (external world). Roughly speaking, they treat of physical objects: bodies, fluids, etc. I am not thinking in particular of the laws of the natural sciences, but of any such proposition as "the tulips in our garden are in full bloom", or "Smith will come in any moment". There are on the other hand propositions describing personal experiences, as when the subject in a psychological experiment describes his sense-experiences; say his visual experience, independent of what bodies are actually before his eyes and, *n.b.*, independent also of any processes which might be observed to take place in his retina, his nerves, his brain, or other parts of his body. (That is, independent of both physical and physiological facts.)

At first sight it may appear (but why it should can only become clear later) that here we have two kinds of worlds, worlds built of different materials; a mental world and a physical world. The mental world in fact is liable to be imagined as gaseous, or rather, aethereal. But let me remind you here of the queer role which the gaseous and the aethereal play in philosophy,—when we perceive that a substantive is not used as what in general we should call the name of an object, and when therefore we can't help saying to ourselves that it is the name of an aethereal object. I mean, we already know the idea of 'aethereal objects' as a subterfuge, when we are embarrassed about the grammar of certain words, and when all we know is that they are not used as names for material objects. This is a hint as to how the problem of the two materials, *mind* and *matter*, is going to dissolve.

It seems to us sometimes as though the phenomena of personal experience were in a way phenomena in the upper strata of the atmosphere as opposed to the material phenomena which happen on the ground. There are views according to which these phenomena in the upper strata arise when the material phenomena reach a certain degree of complexity. E.g., that the mental phenomena, sense experience, volition, etc., emerge when a type of animal body of a certain complexity has been evolved. There seems to be some obvious truth in this, for the amoeba certainly doesn't speak or write or discuss, whereas we do. On the other hand the problem here arises which could be expressed by the question: "Is it possible for a machine to think?" (whether the action of this machine can be described and predicted by the laws of physics or, possibly, only by laws of a different kind applying to the behaviour of organisms). And the trouble which is expressed in this question is not really that we don't yet know a machine which could do the job. The question is not analogous to that which someone might have asked a hundred years ago: "Can a machine liquefy a gas?" The trouble is rather that the sentence, "A machine thinks (perceives, wishes)": seems somehow nonsensical. It is as though we had asked "Has the number 3 a colour?" ("What colour could it be, as it obviously has none of the colours known to us?") For in one aspect of the matter, personal experience, far from being the *product* of physical, chemical, physiological processes, seems to be the very *basis* of all that we say with any sense about such processes. Looking at it in this way we are inclined to use our idea of a building-material in yet another misleading way, and to say that the whole world, mental and physical, is made of one material only.

When we look at everything that we know and can say about the world as resting upon personal experience, then what we know seems to lose a good deal of its value, reliability, and solidity. We are then inclined to say that it is all "subjective"; and "subjective" is used derogatorily, as when we say that an opinion is *merely* subjective, a matter of taste. Now, that this aspect should seem to shake the authority of experience and knowledge points to the fact that here our language is tempting us to draw some misleading analogy. This should remind us of the case when the popular scientist appeared to have shown us that the floor which we stand on is not really solid because it is made up of electrons.

We are up against trouble caused by our way of expression.

Another such trouble, closely akin, is expressed in the sentence: "I can only know that *I* have personal experiences, not that anyone else has".—Shall we then call it an unnecessary hypothesis that anyone else has personal experiences?—But is it an hypothesis at all? For how can I even make the hypothesis if it transcends all possible experience? How could such a hypothesis be backed by meaning? (Is it not like paper money, not backed by gold?)—It doesn't help if anyone tells us that, though we don't know whether the other person has pains, we certainly believe it when, for instance, we pity him. Certainly we shouldn't pity him if we didn't believe that he had pains; but is this a philosophical, a metaphysical belief? Does a realist pity me more than an idealist or a solipsist?—In fact the solipsist asks: "How *can* we believe that the other has pain; what does it mean to believe this? How can the expression of such a supposition make sense?"

Now the answer of the common-sense philosopher—and that, *n.b.*, is not the common-sense man, who is as far from realism as from idealism—the answer of the common-sense philosopher is that surely there is no difficulty in the idea of supposing, thinking, imagining that someone else has what I have. But the trouble with the realist is always that he does not solve but skip the difficulties which his adversaries see, though they too don't succeed in solving them. The realist answer, for us, just brings out the difficulty; for who argues like this overlooks the difference between different usages of the words "to have", "to imagine". "A has a gold tooth" means that the tooth is in A's mouth. This may account for the fact that I am not able to see it. Now the case of his toothache, of which I say that I am not able to feel it because it is in his mouth, is not analogous to the case of the gold tooth. It is the apparent analogy, and again the lack of analogy, between these cases which causes our trouble. And it is this troublesome feature in our grammar which the realist does not notice. It is conceivable that I feel pain in a tooth in another man's mouth; and the man who says that he cannot feel the other's toothache is not denying *this*. The grammatical difficulty which we are in we shall only see clearly if we get familiar with the idea of feeling pain in another person's body. For otherwise, in puzzling about this problem, we shall be liable to confuse our metaphysical proposition "I can't feel his pain" with the experiential proposition, "We can't have (haven't as a rule) pains in another person's tooth". In this proposition the word "can't" is used in the same way as in the proposition "An iron nail can't scratch glass". (We could write this in the form "experience teaches that an iron nail *doesn't* scratch glass", thus doing away with the "can't".) In order to see that it is conceivable that one person should have pain in another person's body, one must examine what sort of facts we call criteria for a pain being in a certain place. It is easy to imagine the following case: When I see my hands I am not always aware of their connection with the rest of my body. That is to say, I often see my hand moving but don't see the arm which connects it to my torso. Nor do I necessarily, at the time, check up on the arm's existence in any other way. Therefore the hand may, for all I know, be connected to the body of a man standing beside me (or, of course, not to a human body at all). Suppose I feel a pain which on the evidence of the pain alone, e.g., with closed eyes, I should call a pain in my left hand. Someone asks me to touch the painful spot with my right hand. I do so and looking round perceive that I am touching my neighbour's hand (meaning the hand connected to my neighbour's torso).

Ask yourself: How do we know where to point to when we are asked to point to the painful spot? Can this sort of pointing be compared with pointing to a black spot on a sheet of paper when someone says: "Point to the black spot on this sheet"? Suppose someone said "You point to this spot because you know before you point that the pains are there"; ask yourself "What does it mean to *know* that the pains are there?" The word "there" refers to a locality;—but in what space, i.e., a 'locality' in what sense? Do we know the place of pain in Euclidean space, so that when we know where we have pains we know how far away from two of the walls of this room, and from the floor? When I have pain in the tip of my finger and touch my tooth with it, is my pain now both a toothache and a pain in my finger? Certainly, in one sense the pain can be said to be located on the tooth. Is the reason why in this case it is wrong to say I have toothache, that in order to be in the tooth the pain should be one sixteenth of an inch away from the tip of my finger? Remember that the word "where" can refer to localities in many different senses. (Many different grammatical games, resembling each other *more* or *less*, are played with this word. Think of the different uses of the numeral "1".) I may know where a thing is and then point to it by virtue of that knowledge. The knowledge tells me where to point to. We here conceived this knowledge as the condition for deliberately pointing to the object. Thus one can say: "I can point to the spot you mean because I see it", "I can direct you to the place because I know where it is; first turning to the right, etc." Now one is inclined to say "I must know where a thing is before I can point to it". Perhaps you will feel less happy about saying: "I must know where a thing is before I can look at it". Sometimes of course it is correct to say this. But we are tempted to think that there is one particular psychical state or event, the knowledge of the place, which must precede every deliberate act of pointing, moving towards, etc. Think of the analogous case: "One can only obey an order after having understood it".

If I point to the painful spot on my arm, in what sense can I be said to have known where the pain was before I pointed to the place? Before I pointed I could have said "The pain is in my left arm". Supposing my arm had been covered with a meshwork of lines numbered in such a way that I could refer to any place on its surface. Was it necessary that I should have been able to describe the painful spot by means of these co-ordinates before I could point to it? What I wish to say is that the act of pointing *determines* a place of pain. This act of pointing, by the way, is not to be confused with that of finding the painful spot by probing. In fact the two may lead to different results.

An innumerable variety of cases can be thought of in which we should say that someone has pains in another person's body; or, say, in a piece of furniture, or in any empty spot. Of course we mustn't forget that a pain in a particular part of our body, e.g., in an upper tooth, has a peculiar tactile and kinaesthetic neighbourhood. Moving our hand upward a little distance we touch our eye; and the word "little distance" here refers to tactile distance or kinaesthetic distance, or both. (It is easy to imagine tactile and kinaesthetic distances correlated in ways different from the usual. The distance from our mouth to our eye might seem very great 'to the muscles of our arm' when we move our finger from the mouth to the eye. Think how large you imagine the cavity in your tooth when the dentist is drilling and probing it.)

When I said that if we moved our hand upward a little, we touch our eye, I was referring to tactile evidence only. That is, the criterion for my finger touching my eye was to be only that I had the particular feeling which would have made me say that I was touching my eye, even if I had no visual evidence for it, and even if, on looking into a mirror, I saw my finger not touching my eye, but, say, my forehead. Just as the 'little distance' I referred to was a tactile or kinaesthetic one, so also the places of which I said, "they lie a little distance apart" were tactile places. To say that my finger in tactile and kinaesthetic space moves from my tooth to my eye then means that I have those tactile and kinaesthetic experiences which we normally have when we say "my finger moves from my tooth to my eye". But what we regard as evidence for this latter proposition is, as we all know, by no means only tactile and kinaesthetic. In fact if I had the tactile and kinaes-thetic sensations referred to, I might still deny the proposition "my finger moves etc. . . ." because of what I saw. That proposition is a proposition about physical objects. (And now don't think that the expression "physical objects" is meant to distin-guish one kind of object from another.) The grammar of proposi-tions which we call propositions about physical objects admits of a variety of evidences for every such proposition. It characterizes the grammar of the proposition "my finger moves, etc." that I regard the propositions "I see it move", "I feel it move", "He sees it move", "He tells me that it moves", etc. as evidences for it. Now if I say "I see my hand move", this at first sight seems to presup-pose that I agree with the proposition "my hand moves". But if I regard the proposition "I see my hand move" as one of the evi-dences for the proposition "my hand moves", the truth of the latter is, of course, not presupposed in the truth of the former. One might therefore suggest the expression "It looks as though my hand were moving" instead of "I see my hand moving". But this expression, although it indicates that my hand may appear to be moving without really moving, might still suggest that af-ter all there must be a hand in order that it should appear to be moving; whereas we could easily imagine cases in which the proposition describing the visual evidence is true and at the same time other evidences make us say that I have no hand. Our ordi-nary way of expression obscures this. We are handicapped in or-dinary language by having to describe, say, a tactile sensation by means of terms for physical objects such as the word "eye", "fin-ger", etc., when what we want to say does not entail the existence of an eye or finger, etc., We have to use a roundabout description of our sensations. This of course does not mean that ordinary language is insufficient for our special purposes, but that it is slightly cumbrous and sometimes misleading. The reason for this peculiarity of our language is of course the regular coincidence of certain sense experiences. Thus when I feel my arm moving I mostly also can see it moving. And if I touch it with my hand, also that hand feels the motion, etc. (The man whose foot has been amputated will describe a particular pain as pain in his foot.) We feel in such cases a strong need for such an expression as: "a sensation travels from my tactual cheek to my tactual eye". I said all this because, if you are aware of the tactual and kinaesthetic environment of a pain, you may find a difficulty in imagining that one could have toothache anywhere else than in one's own teeth. But if we imagine such a case, this simply means that we imagine a correlation between visual, tactual, kinaesthetic, etc., experiences different from the ordinary correlation. Thus we can imagine a person having the sensation of toothache plus those tactual and kinaesthetic experiences which are normally bound up with seeing his hand travelling from his tooth to his nose, to his eyes, etc., but correlated to the visual experience of his hand moving to those places in another person's face. Or again, we can imagine a person having the kinaesthetic sensation of moving his hand, and the tactual sensation, in his fingers and face, of his fingers moving over his face, whereas his kinaesthetic and visual sensations should have to be described as those of his fingers moving over his knee. If we had a sensation of toothache plus certain tactual and kinaesthetic sensations usually characteristic of touching the painful tooth and neighbouring parts of our face, and if these sensations were accompanied by seeing my hand touch, and move about on, the edge of my table, we should feel doubtful whether to call this experience an experience of toothache in the table or not. If, on the other hand, the tactual and kinaesthetic sensations described were correlated to the visual experience of seeing my hand touch a tooth and other parts of the face of another person, there is no doubt that I would call this experience "toothache in another person's tooth".

I said that the man who contended that it was impossible to feel the other person's pain did not thereby wish to deny that one person could feel pain in another person's body. In fact, he would have said: "I may have toothache in another man's tooth, but not *his* toothache".

Thus the propositions "A has a gold tooth" and "A has toothache" are not used analogously. They differ in their grammar where at first sight they might not seem to differ.

As to the use of the word "imagine"—one might say: "Surely there is quite a definite act of imagining the other person to have pain". Of course we don't deny this, or any other statement about facts. But let us see: If we make an image of the other person's pain, do we apply it in the same way in which we apply the image, say, of a black eye, when we imagine the other person having one? Let us again replace imagining, in the ordinary sense, by making a painted image. (This could quite well be *the* way certain beings did their imagining.) Then let a man imagine in this way that A has a black eye. A very important application of this picture will be comparing it with the real eye to see if the picture is correct. When we vividly imagine that someone suffers pain, there often enters in our image what one might call a shadow of a pain felt in the locality corresponding to that in which we say his pain is felt. But the sense in which an image is an image is determined by the way in which it is compared with reality. This we might call the method of projection. Now think of comparing an image of A's toothache with his toothache. How would you compare them? If you say, you compare them 'indirectly' via his bodily behaviour, I answer that this means you *don't* compare them as you compare the picture of his behaviour with his behaviour.

Again, when you say, "I grant you that you can't *know* when A has pain, you can only conjecture it", you don't see the difficulty which lies in the different uses of the words "conjecturing" and "knowing". What sort of impossibility were you referring to when you said you *couldn't* know? Weren't you thinking of a case analogous to that when one couldn't know whether the other man had a gold tooth in his mouth because he had his mouth shut? Here what you didn't know you could nevertheless imagine knowing; it made sense to say that you saw that tooth although you didn't see it; or rather, it makes sense to say that you don't see his tooth and therefore it also makes sense to say that you do. When on the other hand, you granted me that a man can't *know* whether the other person has pain, you do not wish to say that as a matter of fact people didn't know, but that it made no sense to say they knew (and therefore no sense to say they don't know). If therefore in this case you use the term "conjecture" or "believe", you don't use it as opposed to "know". That is, you did not state that knowing was a goal which you could not reach, and that you have to be contented with conjecturing; rather, there is no goal in this game. Just as when one says "You can't count through the whole series of cardinal numbers", one doesn't state a fact about human frailty but about a convention which we have made. Our statement is not comparable, though always falsely compared, with such a one as "it is impossible for a human being to swim across the Atlantic"; but it *is* analogous to a statement like "there is no goal in an endurance race". And this is one of the things which the person feels dimly who is not satisfied with the explanation that though you can't know . . . you can conjecture. . . .

If we are angry with someone for going out on a cold day with a cold in his head, we sometimes say: "I won't feel your cold". And this can mean: "I don't suffer when you catch a cold". This is a proposition taught by experience. For we could imagine a, so to speak, wireless connection between the two bodies which made one person feel pain in his head when the other had exposed his to the cold air. One might in this case argue that the pains are mine because they are felt in my head; but suppose I and someone else had a part of our bodies in common, say a hand. Imagine the nerves and tendons of my arm and A's connected to this hand by an operation. Now imagine the hand stung by a wasp. Both of us cry, contort our faces, give the same description of the pain, etc. Now are we to say we have the same pain or different ones? If in such a case you say: "We feel pain in the same place, in the same body, our descriptions tally, but still my pain can't be his", I suppose as a reason you will be inclined to say: "because my pain is my pain and his pain is his pain". And here you are making a grammatical statement about the use of such a phrase as "the same pain". You say that you don't wish to apply the phrase, "he has got my pain" or "we both have the same pain", and instead, perhaps, you will apply such a phrase as "his pain is exactly like mine". (It would be no argument to say that the two couldn't have the same pain because one might anaesthetize or kill one of them while the other still felt pain.) Of course, if we exclude the phrase "I have his toothache" from our language, we thereby also exclude "I have (or feel) *my* toothache". Another form of our metaphysical statement is this: "A man's sense data are private to himself". And this way of expressing it is even more misleading because it looks still more like an experiential proposition; the philosopher who says this may well think that he is expressing a kind of scientific truth.

We use the phrase "two books have the same colour", but we could perfectly well say: "They can't have the *same* colour, because, after all, this book has its own colour, and the other book has its own colour too". This also would be stating a grammatical rule—a rule, incidentally, not in accordance with our ordinary usage. The reason why one should think of these two different usages at all is this: We compare the case of sense data with that of physical bodies, in which case we make a distinction between: "this is the same chair that I saw an hour ago" and "this is not the same chair, but one exactly like the other". Here it makes sense to say, and it is an experiential proposition: "A and B couldn't have seen the same chair, for A was in London and B in Cambridge; they saw two chairs exactly alike". (Here it will be useful if you consider the different criteria for what we call the "identity of these objects". How do we apply the statements: "This is the same day . . .", "This is the same word . . .", "This is the same occasion . . .", etc.?)

What we did in these discussions was what we always do when we meet the word "can" in a metaphysical proposition. We show that this proposition hides a grammatical rule. That is to say, we destroy the outward similarity between a metaphysical proposition and an experiential one, and we try to find the form of expression which fulfils a certain craving of the metaphysician which our ordinary language does not fulfil and which, as long as it isn't fulfilled, produces the metaphysical puzzlement. Again, when in a metaphysical sense I say "I *must* always know when I have pain", this simply makes the word "know" redundant; and instead of "I know that I have pain", I can simply say "I have pain". The matter is different, of course, if we give the phrase "unconscious pain" sense by fixing experiential criteria for the case in which a man has pain and doesn't know it, and if then we say (rightly or wrongly) that as a matter of fact nobody has ever had pains which he didn't know of.

When we say "I can't feel his pain", the idea of an insurmountable barrier suggests itself to us. Let us think straight away of a similar case: "The colours green and blue can't be in the same place simultaneously". Here the picture of physical impossibility which suggests itself is, perhaps, not that of a barrier; rather we feel that the two colours are in each other's way. What is the origin of this idea?—We say three people can't sit side by side on this bench; they have no room. Now the case of the colours is not analogous to this; but it is somewhat analogous to saying: "3 × 18 inches won't go into 3 feet". This is a grammatical rule and states a logical impossibility. The proposition "three men can't sit side by side on a bench a yard long" states a physical impossibility; and this example shows clearly why the two impossibilities are confused. (Compare the proposition "He is 6 inches taller than I" with "6 foot is 6 inches longer than 5 foot 6". These propositions are of utterly different kinds, but look exactly alike.) The reason why in these cases the idea of physical impossibility suggests itself to us is that on the one hand we decide against using a particular form of expression, on the other hand we are strongly tempted to use it, since (*a*) it sounds English, or German, etc., all right, and (*b*) there are closely similar forms of expression used in other departments of our language. We have decided against using the phrase "They are in the same place"; on the other hand this phrase strongly recommends itself to us through the analogy with other phrases, so that, in a sense, we have to turn this form of expression out by force. And this is why we seem to ourselves to be rejecting a universally false proposition. We make a picture like that of the two colours being in each other's way, or that of a barrier which doesn't allow one person to come closer to another's experience than to the point of observing his behaviour; but on looking closer we find that we can't apply the picture which we have made.

Our wavering between logical and physical impossibility makes us make such statements as this: "If what I feel is always *my* pain only, what can the supposition mean that someone else has pain?" The thing to do in such cases is always to look how the words in question *are actually used in our language.* We are in all such cases thinking of a use different from that which our ordinary language makes of the words. Or a use, on the other hand, which just then for some reason strongly recommends itself to us. When something seems queer about the grammar of our words, it is because we are alternately tempted to use a word in several different ways. And it is particularly difficult to discover that an assertion which the metaphysician makes expresses discontentment with our grammar when the words of this assertion can also be used to state a fact of experience. Thus when he says "only my pain is real pain", this sentence might mean that the other people are only pretending. And when he says "this tree doesn't exist when nobody sees it", this might mean: "this tree vanishes when we turn our backs to it". The man who says "only my pain is real", doesn't mean to say that he has found out by the common criteria—the criteria, i.e., which give our words their common meanings—that the others who said they had pains were cheating. But what he rebels against is the use of *this* expression in connection with *these* criteria. That is, he objects to using this word in the particular way in which it is commonly used. On the other hand, he is not aware that he is objecting to a convention. He sees a way of dividing the country different from the one used on the ordinary map. He feels tempted, say, to use the name "Devonshire" not for the county with its conventional boundary, but for a region differently bounded. He could express this by saying: "Isn't it absurd to make *this* a county, to draw the boundaries *here*?" But what he says is: "The *real* Devonshire is this". We could answer: "What you want is only a new notation, and by a new notation no facts of geography are changed". It is true, however, that we may be irresistibly attracted or repelled by a notation. (We easily forget how much a notation, a form of expression, may mean to us, and that changing it isn't always as easy as it often is in mathematics or in the sciences. A change of clothes or of names may mean very little and it may mean a great deal.)

I shall try to elucidate the problem discussed by realists, idealists, and solipsists by showing you a problem closely related to it. It is this: "Can we have unconscious thoughts, unconscious feelings, etc.?" The idea of there being unconscious thoughts has revolted many people. Others again have said that these were wrong in supposing that there could only be conscious thoughts, and that psychoanalysis had discovered unconscious ones. The objectors to unconscious thought did not see that they were not objecting to the newly discovered psychological reactions, but to the way in which they were described. The psychoanalysts on the other hand were misled by their own way of expression into thinking that they had done more than discover new psychological reactions; that they had, in a sense, discovered conscious thoughts which were unconscious. The first could have stated their objection by saying "We don't wish to use the phrase 'unconscious thoughts'; we wish to reserve the word 'thought' for what you call 'conscious thoughts'". They state their case wrongly when they say: "There can only be conscious thoughts and no unconscious ones". For if they don't wish to talk of "unconscious thought" they should not use the phrase "conscious thought", either.

But is it not right to say that in any case the person who talks both of conscious and unconscious thoughts thereby uses the word "thoughts" in two different ways?—Do we use a hammer in two different ways when we hit a nail with it and, on the other hand, drive a peg into a hole? And do we use it in two different ways or in the same way when we drive this peg into this hole and, on the other hand, another peg into another hole? Or should we only call it different uses when in one case we drive something into something and in the other, say, we smash something? Or is this all using the hammer in one way and is it to be called a different way only when we use the hammer as a paper weight?—In which cases are we to say that a word is used in two different ways and in which that it is used in one way? To say that a word is used in two (or more) different ways does in itself not yet give us any idea about its use. It only specifies a way of looking at this usage by providing a schema for its description with two (or more) subdivisions. It is all right to say: "I do *two* things with this hammer: I drive a nail into this board and one into that board". But I could also have said: "I am doing only one thing with this hammer; I am driving a nail into this board and one into that board". There can be two kinds of discussions as to whether a word is used in one way or in two ways: (*a*) Two people may discuss whether the English word "cleave" is only used for chopping up something or also for joining things together. This is a discussion about the facts of a certain actual usage. (*b*) They may discuss whether the word "altus", standing for both "deep" and "high", is *thereby* used in two different ways. This question is analogous to the question whether the word "thought" is used in two ways or in one when we talk of conscious and unconscious thought. The man who says "surely, these are two different usages" has already decided to use a two-way schema, and what he said expressed this decision.

Now when the solipsist says that only his own experiences are real, it is no use answering him: "Why do you tell us this if you don't believe that we really hear it?" Or anyhow, if we give him this answer, we mustn't believe that we have answered his difficulty. There is no common sense answer to a philosophical problem. One can defend commonsense against the attacks of philosophers only by solving their puzzles, i.e., by curing them of the temptation to attack common sense; not by restating the views of common sense. A philosopher is not a man out of his senses, a man who doesn't see what everybody sees; nor on the other hand is his disagreement with common sense that of the scientist disagreeing with the coarse views of the man in the street. That is, his disagreement is not founded on a more subtle knowledge of fact. We therefore have to look round for the *source* of his puzzlement. And we find that there is puzzlement and mental discomfort, not only when our curiosity about certain facts is not satisfied or when we can't find a law of nature fitting in with all our experience, but also when a notation dissatisfies us—perhaps because of various associations which it calls up. Our ordinary language, which of all possible notations is the one which pervades all our life, holds our mind rigidly in one position, as it were, and in this position sometimes it feels cramped, having a desire for other positions as well. Thus we sometimes wish for a notation which stresses a difference more strongly, makes it more obvious, than ordinary language does, or one which in a particular case uses more closely similar forms of expression than our ordinary language. Our mental cramp is loosened when we are shown the have only learnt the word 'seeing' to mean what *I* do". Of course this is just not true, for I have definitely learned a different and much more complicated use of the word "to see" than I here profess. Let us make clear the tendency which guided me when I did so, by an example from a slightly different sphere: Consider this argument: "How can we wish that this paper were red if it isn't red? Doesn't this mean that I wish that which doesn't exist at all? Therefore my wish can only contain something *similar* to the paper's being red. Oughtn't we therefore to use a different word instead of 'red' when we talk of wishing that something were red? The imagery of the wish surely shows us something less definite, something hazier, than the reality of the paper being red. I should therefore say, instead of 'I wish this paper were red', something like 'I wish a pale red for this paper'". But if in the usual way of speaking he had said, "I wish a pale red for this paper," we should, in order to fulfil his wish, have painted it a pale red—and this wasn't what he wished. On the other hand there is no objection to adopting the form of expression which he suggests as long as we know that he uses the phrase "I wish a pale *x* for this paper", always to mean what ordinarily we express by "I wish this paper had the colour *x*". What he said really recommended his notation, in the sense in which a notation can be recommended. But he did not tell us a new truth and did not show us that what we said before was false. (All this connects our present problem with the problem of negation. I will only give you a hint, by saying that a notation would be possible in which, to put it roughly, a quality had always two names, one for the case when something is said to have it, the other for the case when something is said not to have it. The negation of "This paper is red" could then be, say, "This paper is not rode". Such a notation would actually fulfil some of the wishes which are denied us by our ordinary language and which sometimes produce a cramp of philosophical puzzlement about the idea of negation.)

The difficulty which we express by saying "I can't know what he sees when he (truthfully) says that he sees a blue patch" arises from the idea that "knowing what he sees" means: "seeing that which he also sees"; not, however, in the sense in which we do so when we both have the same object before our eyes: but in the sense in which the object seen would be an object, say, in his head, or in *him*. The idea is that the same object may be before his eyes and mine, but that I can't stick my head into his (or my mind into his, which comes to the same) so that the *real* and *immediate* object of his vision becomes the real and immediate object of my vision too. By "I don't know what he sees" we really mean "I don't know what he looks at", where 'what he looks at' is hidden and he can't show it to me; it is *before his mind's eye*. Therefore, in order to get rid of this puzzle, examine the grammatical difference between the statements "I don't know what he sees" and "I don't know what he looks at", as they are actually used in our language.

Sometimes the most satisfying expression of our solipsism seems to be this: "When anything is seen (really *seen*), it is always I who see it".

What should strike us about this expression is the phrase "always I". Always *who*?—For, queer enough, I don't mean: "always L. W." This leads us to considering the criteria for the identity of a person. Under what circumstances do we say: "This is the same person whom I saw an hour ago"? Our actual use of the phrase "the same person" and of the name of a person is based on the fact that many characteristics which we use as the criteria for identity coincide in the vast majority of cases. I am as a rule recognized by the appearance of my body. My body changes its appearance only gradually and comparatively little, and likewise my voice, characteristic habits, etc. only change slowly and within a narrow range. We are inclined to use personal names in the way we do, only as a consequence of these facts. This can best be seen by imagining unreal cases which show us what different 'geometries' we would be inclined to use if facts were different. Imagine, e.g., that all human bodies which exist looked alike, that on the other hand, different sets of characteristics seemed, as it were, to change their habitation among these bodies. Such a set of characteristics might be, say, mildness, together with a high pitched voice, and slow movements, or a choleric temperament, a deep voice, and jerky movements, and such like. Under such circumstances, although it would be possible to give the bodies names, we should perhaps be as little inclined to do so as we are to give names to the chairs of our dining-room set. On the other hand, it might be useful to give names to the sets of characteristics, and the use of these names would now *roughly* correspond to the personal names in our present language.

Or imagine that it were usual for human beings to have two characters, in this way: People's shape, size and characteristics of behaviour periodically undergo a complete change. It is the usual thing for a man to have two such states, and he lapses suddenly from one into the other. It is very likely that in such a society we should be inclined to christen every man with two names, and perhaps to talk of the pair of persons in his body. Now were Dr. Jekyll and Mr. Hyde two persons or were they the same person who merely changed? We can say whichever we like. We are not forced to talk of a double personality.

There are many uses of the word "personality" which we may feel inclined to adopt, all more or less akin. The same applies when we define the identity of a person by means of his memories. Imagine a man whose memories on the even days of his life comprise the events of all these days, skipping entirely what happened on the odd days. On the other hand, he remembers on an odd day what happened on previous odd days, but his memory then skips the even days without a feeling of discontinuity. If we like we can also assume that he has alternating appearances and characteristics on odd and even days. Are we bound to say that here two persons are inhabiting the same body? That is, is it right identity of physical objects? (To understand this sentence, you must remember that the grammar of words of which we say that they stand for physical objects is characterized by the way in which we use the phrase "the *same* so-and-so", or "the identical so-and-so", where "so-and-so" designates the physical object.) We said before that he did not wish to point to a particular physical object at all. The idea that he had made a significant statement arose from a confusion corresponding to the confusion between what we shall call "the geometrical eye" and "the physical eye". I will indicate the use of these terms: If a man tries to obey the order "Point to your eye", he may do many different things, and there are many different criteria which he will accept for having pointed to his eye. If these criteria, as they usually do, coincide, I may use them alternately and in different combinations to show me that I have touched my eye. If they don't coincide, I shall have to distinguish between different senses of the phrase "I touch my eye" or "I move my finger towards my eye". If, e.g., my eyes are shut, I can still have the characteristic kinaesthetic experience in my arm which I should call the kinaesthetic experience of raising my hand to my eye. That I had succeeded in doing so, I shall recognize by the peculiar tactile sensation of touching my eye. But if my eye were behind a glass plate fastened in such a way that it prevented me from exerting a pressure on my eye with my finger, there would still be a criterion of muscular sensation which would make me say that now my finger was in front of my eye. As to visual criteria, there are two I can adopt. There is the ordinary experience of seeing my hand rise and come towards my eye, and this experience, of course, is different from seeing two things meet, say, two finger tips. On the other hand, I can use as a criterion for my finger moving towards my eye, what I see when I look into a mirror and see my finger nearing my eye. If that place on my body which, we say, 'sees' is to be determined by moving my finger towards my eye, according to the second criterion, then it is conceivable that I may see with what according to other criteria is the tip of my nose, or places on my forehead; or I might in this way point to a place lying outside my body. If I wish a person to point to his eye (or his eyes) according to the second criterion *alone*, I shall express my wish by saying: "Point to your geometrical eye (or eyes)". The grammar of the word "geometrical eye" stands in the same relation to the grammar of the word "physical eye" as the grammar of the expression "the visual sense datum of a tree" to the grammar of the expression "the physical tree". In either case it confuses everything to say "the one is a *different kind* of object from the other"; for those who say that a sense datum is a different kind of object from a physical object misunderstand the grammar of the word "kind", just as those who say that a number is a different kind of object from a numeral. They think they are making such a statement as "A railway train, a railway station, and a railway car are different kinds of objects", whereas their statement is analogous to "A railway train, a railway accident, and a railway law are different kinds of objects".

What tempted me to say "it is always I who see when anything is seen", I could also have yielded to by saying: "whenever anything is seen, it is *this* which is seen", accompanying the word "this" by a gesture embracing my visual field (but not meaning by "this" the particular objects which I happen to see at the moment). One might say, "I am pointing at the visual field as such, not at anything in it". And this only serves to bring out the senselessness of the former expression.

Let us then discard the "always" in our expression. Then I can still express my solipsism by saying, "Only what *I* see (or: see now) is really seen". And here I am tempted to say: "Although by the word 'I' I don't mean L. W., it will do if the others understand 'I' to mean L. W., if just now I am in fact L. W." I could also express my claim by saying: "I am the vessel of life"; but mark, it is essential that everyone to whom I say this should be unable to understand me. It is essential that the other should not be able to understand 'what *I* really *mean*', though in practice he might do what I wish by conceding to me an exceptional position in his notation. But I wish it to be *logically* impossible that he should understand me, that is to say, it should be meaningless, not false, to say that he understands me. Thus my expression is one of the many which is used on various occasions by philosophers and supposed to convey something to the person who says it, though essentially incapable of conveying anything to anyone else. Now if for an expression to convey a meaning means to be accompanied by or to produce certain experiences, our expression may have all sorts of meanings, and I don't wish to say anything about them. But we are, as a matter of fact, misled into thinking that our expression has a meaning in the sense in which a non-metaphysical expression has; for we wrongly compare our case with one in which the other person can't understand what we say because he lacks a certain information. (This remark can only become clear if we understand the connection between grammar and sense and nonsense.)

The meaning of a phrase for us is characterized by the use we make of it. The meaning is not a mental accompaniment to the expression. Therefore the phrase "I think I mean something by it", or "I'm sure I mean something by it", which we so often hear in philosophical discussions to justify the use of an expression is for us no justification at all. We ask: "*What* do you mean?", i.e., "How do you use this expression?" If someone taught me the word "**bench**" and said that he sometimes or always put a stroke over it thus: "$\overline{\text{bench}}$", and that this meant something to him, I should say: "I don't know what sort of idea you associate with this stroke, but it doesn't interest me unless you show me that there is a use for the stroke in the kind of calculus in which you wish to use the word 'bench'".—I want to play chess, and a man gives the white king a paper crown, leaving the use of the piece unaltered, but telling me that the crown has a meaning to him in the game, which he can't express by rules. I say: "as long my mind, but could only point to it via my body.) There is nothing wrong in suggesting that the others should give me an exceptional place in their notation; but the justification which I wish to give for it: that this body is now the seat of that which really lives—is senseless. For admittedly this is not to state anything which in the ordinary sense is a matter of experience. (And don't think that it is an experiential proposition which only I can know because only I am in the position to have the particular experience.) Now the idea that the real I lives in my body is connected with the peculiar grammar of the word "I", and the misunderstandings this grammar is liable to give rise to. There are two different cases in the use of the word "I" (or "my") which I might call "the use as object" and "the use as subject". Examples of the first kind of use are these: "My arm is broken", "I have grown six inches", "I have a bump on my forehead", "The wind blows my hair about". Examples of the second kind are: "*I* see so-and-so", "*I* hear so-and-so", "*I* try to lift my arm", "*I* think it will rain", "*I* have toothache". One can point to the difference between these two categories by saying: The cases of the first category involve the recognition of a particular person, and there is in these cases the possibility of an error, or as I should rather put it: The possibility of an error has been provided for. The possibility of failing to score has been provided for in a pin game. On the other hand, it is not one of the hazards of the game that the balls should fail to come up if I have put a penny in the slot. It is possible that, say in an accident, I should feel a pain in my arm, see a broken arm at my side, and think it is mine, when really it is my neighbour's. And I could, looking into a mirror, mistake a bump on his forehead for one on mine. On the other hand, there is no question of recognizing a person when I say I have toothache. To ask "are you sure that it's *you* who have pains?" would be nonsensical. Now, when in this case no error is possible, it is because the move which we might be inclined to think of as an error, a 'bad move', is no move of the game at all.

(We distinguish in chess between good and bad moves, and we call it a mistake if we expose the queen to a bishop. But it is no mistake to promote a pawn to a king.) And now this way of stating our idea suggests itself: that it is as impossible that in making the statement "I have toothache" I should have mistaken another person for myself, as it is to moan with pain by mistake, having mistaken someone else for me. To say, "I have pain" is no more a statement *about* a particular person than moaning is. "But surely the word 'I' in the mouth of a man refers to the man who says it; it points to himself; and very often a man who says it actually points to himself with his finger". But it was quite superfluous to point to himself. He might just as well only have raised his hand. It would be wrong to say that when someone points to the sun with his hand, he is pointing both to the sun and himself because it is *he* who points; on the other hand, he may by pointing attract attention both to the sun and to himself.

The word "I" does not mean the same as "L. W." even if I am L. W., nor does it mean the same as the expression "the person who is now speaking". But that doesn't mean: that "L. W." and "I" mean different things. All it means is that these words are different instruments in our language.

Think of words as instruments characterized by their use, and then think of the use of a hammer, the use of a chisel, the use of a square, of a glue pot, and of the glue. (Also, all that we say here can be understood only if one understands that a great variety of games is played with the sentences of our language: Giving and obeying orders; asking questions and answering them; describing an event; telling a fictitious story; telling a joke; describing an immediate experience; making conjectures about events in the physical world; making scientific hypotheses and theories; greeting someone, etc., etc.) The mouth which says "I" or the hand which is raised to indicate that it is I who wish to speak, or I who have toothache, does not thereby point to anything. If, on the other hand, I wish to indicate the *place* of my pain, I point. And here again remember the difference between pointing to the painful spot without being led by the eye and on the other hand pointing to a scar on my body after looking for it. ("That's where I was vaccinated".)—The man who cries out with pain, or says that he has pain, *doesn't choose the mouth which says it*.

All this comes to saying that the person of whom we say "he has pain" is, by the rules of the game, the person who cries, contorts his face, etc. The place of the pain—as we have said—may be in another person's body. If, in saying "I", I point to my own body, I model the use of the word "I" on that of the demonstrative "this person" or "he". (This way of making the two expressions similar is somewhat analogous to that which one sometimes adopts in mathematics, say in the proof that the sum of the three angles of a triangle is 180°.

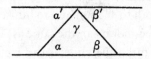

We say "$\alpha = \alpha'$, $\beta = \beta'$, and $\underline{\gamma = \gamma}$". The first two equalities are of an entirely different kind from the third.) In "I have pain", "I" is not a demonstrative pronoun.

Compare the two cases: 1. "How do you know that *he* has pains?"—"Because I hear him moan". 2. "How do you know that you have pains?"—"Because I *feel* them". But "I feel them" means the same as "I have them". Therefore this was no explanation at all. That, however, in my answer I am inclined to stress the word "feel" and not the word "I" indicates that by "I" I don't wish to pick out one person (from amongst different persons).

The difference between the propositions "I have pain" and "he has pain" is not that of "L. W. has pain" and "Smith has pain". Rather, it corresponds to the difference between moaning and saying that someone moans.—"But surely the word 'I' in 'I have pain' serves to distinguish me from other people, because it is by the sign 'I' that I distinguish saying that I have pain from saying that one of the others has". Imagine a language in which, instead of "I found nobody in the room", one said "I found Mr. Nobody in the room". Imagine the philosophical problems which would arise out of such a convention. Some philosophers brought up in this language would probably feel that they didn't like the similarity of the expressions "Mr. Nobody" and "Mr. Smith". When we feel that we wish to abolish the "I" in "I have pain", one may say that we tend to make the verbal expression of pain similar to the expression by moaning.—We are inclined to forget that it is the particular use of a word only which gives the word its meaning. Let us think of our old example for the use of words: Someone is sent to the grocer with a slip of paper with the words "five apples" written on it. The use of the term "in practice" is its meaning. Imagine it were the usual thing that the objects around us carried labels with words on them by means of which our speech referred to the objects. Some of these words would be proper names of the objects, others generic names (like table, chair, etc.), others again, names of colours, names of shapes, etc. That is to say, a label would only have a meaning to us in so far as we made a particular use of it. Now we could easily imagine ourselves to be impressed by merely seeing a label on a thing, and to forget that what makes these labels important is their use. In this way we sometimes believe that we have named something when we make the gesture of pointing and utter words like "This is . . ." (the formula of the ostensive definition). We say we call something "toothache", and think that the word has received a definite function in the dealings we carry out with language when, under certain circumstances, we have pointed to our cheek and said: "This is toothache". (Our idea is that when we point and the other "only knows what we are pointing to" he knows the use of the word. And here we have in mind the special case when 'what we point to' is, say, a person and "to know that I point to" means to see which of the persons present I point to.)

We feel then that in the cases in which "I" is used as subject, we don't use it because we recognize a particular person by his bodily characteristics; and this creates the illusion that we use this word to refer to something bodiless, which, however, has its seat in our body. In fact *this* seems to be the real ego, the one of which it was said, "Cogito ergo sum".—"Is there then no mind, but only a body?" Answer: The word "mind" has meaning, i.e., it has a use in our language; but saying this doesn't yet say what kind of use we make of it.

In fact one may say that what in these investigations we were concerned with was the grammar of those words which describe what are called "mental activities": seeing, hearing, feeling, etc. And this comes to the same as saying that we are concerned with the grammar of 'phrases describing sense data'.

Philosophers say it as a philosophical opinion or conviction that there are sense data. But to say that I believe that there are sense data comes to saying that I *believe* that an object may appear to be before our eyes even when it isn't. Now when one uses the term "sense datum", one should be clear about the peculiarity of its grammar. For the idea in introducing this expression was to model expressions referring to 'appearance' after expressions referring to 'reality'. It was said, e.g., that if two things *seem* to be equal, there *must* be two somethings which *are* equal. Which of course means nothing else but that we have decided to use such an expression as "the appearances of these two things are equal" synonymously with "these two things seem to be equal". Queerly enough, the introduction of this new phraseology has deluded people into thinking that they had discovered new entities, new elements of the structure of the world, as though to say "I believe that there are sense data" were similar to saying "I believe that matter consists of electrons". When we talk of the equality of appearances or sense data, we introduce a new usage of the word "equal". It is possible that the lengths A and B should appear to us to be equal, that B and C should appear to be equal, but that

A and C do not appear to be equal. And in the new notation we shall have to say that though the appearance (sense datum) of A is equal to that of B and the appearance of B equal to that of C, the appearance of A is not equal to the appearance of C; which is all right if you don't mind using "equal" intransitively.

Now the danger we are in when we adopt the sense datum notation is to forget the difference between the grammar of a statement about sense data and the grammar of an outwardly similar statement about physical objects. (From this point one might go on talking about the misunderstandings which find their expression in such sentences as: "We can never see an accurate circle", "All our sense data are vague". Also, this leads to the comparison of the grammar of "position", "motion", and "size" in Euclidean and in visual space. There is, e.g., absolute position, absolute motion and size, in visual space.)

Now we can make use of such an expression as "pointing to the *appearance* of a body" or "pointing to a visual sense datum". Roughly speaking, this sort of pointing comes to the same as sighting, say, along the barrel of a gun. Thus we may point and say: "This is the direction in which I see my image in the mirror". One can also use such an expression as "the appearance, or sense datum, of my finger points to the sense datum of the tree" and similar ones. From these cases of pointing, however, we must distinguish those of pointing in the direction a sound seems to come from, or of pointing to my forehead with closed eyes, etc.

Now when in the solipsistic way I say "*This* is what's really seen", I point before me and it is essential that I point *visually*. If I pointed sideways or behind me—as it were, to things which I don't see—the pointing would in this case be meaningless to me; it would not be pointing in the sense in which I wish to point. But this means that when I point before me saying "this is what's really seen", although I make the gesture of pointing, I don't point to one thing as opposed to another. This is as when travelling in a car and feeling in a hurry, I instinctively press against something in front of me as though I could push the car from inside.

When it makes sense to say "I see this", or "this is seen", pointing to what I see, it also makes *sense* to say "I see this", or "this is seen", pointing to something I *don't* see. When I made my solipsist statement, I pointed, but I robbed the pointing of its sense by inseparably connecting that which points and that to which it points. I constructed a clock with all its wheels, etc., and in the end fastened the dial to the pointer and made it go round with it. And in this way the solipsist's "Only this is really seen" reminds us of a tautology.

Of course one of the reasons why we are tempted to make our pseudo-statement is its similarity with the statement "I only see this", or "this is the region which I see", where I point to certain objects around me, as opposed to others, or in a certain direction in physical space (not in visual space), as opposed to other directions in physical space. And if, pointing in this sense, I say "this is what is really seen", one may answer me: "This is what *you*, L. W., see; but there is no objection to adopting a notation in which what we used to call 'things which L. W. sees' is called 'things really seen'". If, however, I believe that by pointing to that which in my grammar has no neighbour I can convey something to myself (if not to others), I make a mistake similar to that of thinking that the sentence "I am here" makes sense to me (and, by the way, is always true) under conditions different from those very special conditions under which it does make sense. E.g., when my voice and the direction from which I speak is recognized by another person. Again an important case where you can learn that a word has meaning by the particular use we make of it.—We are like people who think that pieces of wood shaped more or less like chess or draught pieces and standing on a chess board make a game, even if nothing has been said as to how they are to be used.

To say "it approaches me" has sense, even when, physically

speaking, nothing approaches my body; and in the same way it makes sense to say, "it is here" or "it has reached me" when nothing has reached my body. And, on the other hand, "I am here" makes sense if my voice is recognized and heard to come from a particular place of common space. In the sentence "it is here" the 'here' was a here in visual space. Roughly speaking, it is the geometrical eye. The sentence "I am here", to make sense, must attract attention to a place in common space. (And there are several ways in which this sentence might be used.) The philosopher who thinks it makes sense to say to himself "I am here" takes the verbal expression from the sentence in which "here" is a place in common space and thinks of "here" as the here in visual space. He therefore really says something like "Here is here".

I could, however, try to express my solipsism in a different way: I imagine that I and others draw pictures or write descriptions of what each of us sees. These descriptions are put before me. I point to the one which I have made and say: "Only this is (or was) really seen". That is, I am tempted to say: "Only this description has reality (visual reality) behind it". The others I might call—"blank descriptions". I could also express myself by saying: "*This* description only was derived from reality; only this was compared with reality". Now it has a clear meaning when we say that this picture or description is a projection, say, of this group of objects—the trees I look at—or that it has been derived from these objects. But we must look into the grammar of such a phrase as "this description is derived from my sense datum". What we are talking about is connected with that peculiar temptation to say: "I never know what the other really means by 'brown', or what he really sees when he (truthfully) says that he sees a brown object".—We could propose to one who says this to use two different words instead of the one word "brown"; one word *for his particular impression*, the other word with that meaning which other people besides himself can understand as well. If he thinks about this proposal he will see that there is something wrong in his conception of the meaning, function, of the word "brown" and others. He looks for a justification of his description where there is none. (Just as in the case when a man believes that the chain of reasons must be endless. Think of the justification by a general formula for performing mathematical operations; and of the question: Does this formula compel us to make use of it in this particular case as we do?) To say "I derive a description from visual reality" can't mean anything analogous to: "I derive a description from what I see here". I may, e.g., see a chart in which a coloured square is correlated to the word "brown", and also a patch of the same colour elsewhere; and I may say: "This chart shows me that I must use the word 'brown' for the description of this patch". This is how I may derive the word which is needed in my description. But it would be meaningless to say that I derive the word "brown" from the particular colour-impression which I receive.

Let us now ask: "Can a human *body* have pain?" One is inclined to say: "How can the body have pain? The body in itself is something dead; a body isn't conscious!" And here again it is as though we looked into the nature of pain and saw that it lies in its nature that a material object can't have it. And it is as though we saw that what has pain must be an entity of a different nature from that of a material object; that, in fact, it must be of a mental nature. But to say that the ego is mental is like saying that the number 3 is of a mental or an immaterial nature, when we recognize that the numeral "3" isn't used as a sign for a physical object.

On the other hand we can perfectly well adopt the expression "this body feels pain", and we shall then, just as usual, tell it to go to the doctor, to lie down, and even to remember that when the last time it had pains they were over in a day. "But wouldn't this form of expression at least be an indirect one?"—Is it using an indirect expression when we say "Write '3' for '*x*' in this formula" instead of "Substitute 3 for *x*"? (Or on the other hand, is the first of these two expressions the only direct one, as some philosophers think?) One expression is no more direct than the other. The meaning of the expression depends entirely on how we go on using it. Let's not imagine the meaning as an occult connection the mind makes between a word and a thing, and that this connection *contains* the whole usage of a word as the seed might be said to contain the tree.

The kernel of our proposition that that which has pains or sees or thinks is of a mental nature is only, that the word "I" in "I have pains" does not denote a particular body, for we can't substitute for "I" a description of a body.

"cube", "brick", "slab", "column". A calls out one of these words, upon which B brings a stone of a certain shape. Let us imagine a society in which this is the only system of language. The child learns this language from the grown-ups by being trained to its use. I am using the word "trained" in a way strictly analogous to that in which we talk of an animal being trained to do certain things. It is done by means of example, reward, punishment, and suchlike. Part of this training is that we point to a building stone, direct the attention of the child towards it, and pronounce a word. I will call this procedure *demonstrative* teaching of words. In the actual use of this language, one man calls out the words as orders, the other acts according to them. But learning and teaching this language will contain this procedure: The child just 'names' things, that is, he pronounces the words of the language when the teacher points to the things. In fact, there will be a still simpler exercise: The child repeats words which the teacher pronounces.

(Note. Objection: The word "brick" in language 1) has not the meaning which it has in *our* language.——This is true if it means that in our language there are usages of the word "brick" different from our usages of this word in language 1). But don't we sometimes use the word "brick!" in just this way? Or should we say that when we use it, it is an elliptical sentence, a shorthand for "Bring me a brick"? Is it right to say that if *we* say "brick!" we *mean* "Bring me a brick"? Why should I translate the expression "brick!" into the expression "Bring me a brick"? And if they are synonymous, why shouldn't I say: If he says "brick!" he means "brick!" . . . ? Or: Why shouldn't he be able to mean just "brick!" if he is able to mean "Bring me a brick", unless you wish to assert that while he says aloud "brick!" he as a matter of fact always says in his mind, to himself, "Bring me a brick"? But what reason could we have to assert this? Suppose someone asked: If a man gives the order, "Bring me a brick", must he mean it as four words, or can't he mean it as one composite word synonymous with the one word "brick!"? One is tempted to answer: He *means* all four words if in his language he uses that sentence in contrast with other sentences in which these words are used, such as, for instance, "Take these two bricks away". But what if I asked "But how is his sentence contrasted with these others? Must he have thought them simultaneously, or shortly before or after, or is it sufficient that he should have one time learnt them, etc.?" When we have asked ourselves this question, it appears that it is irrelevant which of these alternatives is the case. And we are inclined to say that all that is really relevant is that these contrasts should exist in the system of language which he is using, and that they need not in any sense be present in his mind when he utters his sentence. Now compare this conclusion with our original question. When we asked it, we seemed to ask a question about the state of mind of the man who says the sentence, whereas the idea of meaning which we arrived at in the end was not that of a state of mind. We think of the meaning of signs sometimes as states of mind of the man using them, sometimes as the role which these signs are playing in a system of language. The connection between these two ideas is that the mental experiences which accompany the use of a sign undoubtedly are caused by our usage of the sign in a particular system of language. William James speaks of specific feelings accompanying the use of such words as "and", "if", "or". And there is no doubt that at least certain gestures are often connected with such words, as a collecting gesture with "and", and a dismissing gesture with "not". And there obviously are visual and muscular sensations connected with these gestures. On the other hand it is clear enough that these sensations do not accompany every use of the word "not" and "and". If in some language the word "but" meant what "not" means in English, it is clear that we should not compare the meanings of these two words by comparing the sensations which they produce. Ask yourself what means we have of finding out the feelings which they produce in different people and on says, "In one case we point to a shape, in the other we point to a number". The difference becomes obvious and clear only when we contemplate a *complete* example (i.e., the example of a language completely worked out in detail).)

3). Let us introduce a new instrument of communication,—a proper name. This is given to a particular object (a particular building stone) by pointing to it and pronouncing the name. If A calls the name, B brings the object. The demonstrative teaching of a proper name is different again from the demonstrative teaching in the cases 1) and 2).

(Remark: This difference does not lie, however, in the act of pointing and pronouncing the word or in any mental act [meaning?] accompanying it, but in the role which the demonstration [pointing and pronouncing] plays in the whole training and in the use which is made of it in the practice of communication by means of this language. One might think that the difference could be described by saying that in the different cases we point to different kinds of objects. But suppose I point with my hand to a blue jersey. How does pointing to its colour differ from pointing to its shape?—We are inclined to say the difference is that we *mean* something different in the two cases. And 'meaning' here is to be some sort of process taking place while we point. What particularly tempts us to this view is that a man on being asked whether he pointed to the colour or the shape is, at least in most cases, able to answer this and to be certain that his answer is correct. If on the other hand, we look for two such characteristic mental acts as meaning the colour and meaning the shape, etc., we aren't able to find any, or at least none which must always accompany pointing to colour, pointing to shape, respectively. We have only a *rough* idea of what it means to concentrate one's attention on the colour as opposed to the shape, or vice versa. The difference, one might say, does not lie in the act of demonstration, but rather in the surrounding of that act in the use of the language.)

4).   On being ordered, "This slab!", B brings the slab to which A points. On being ordered, "Slab, there!", he carries a slab to the place indicated. Is the word "there" taught demonstratively? Yes and no! When a person is trained in the use of the word "there", the teacher will in training him make the pointing gesture and pronounce the word "there". But should we say that thereby he gives a place the name "there"? Remember that the pointing gesture in this case is part of the practice of communication itself.

(Remark: It has been suggested that such words as "there", "here", "now", "this" are the *real proper names* as opposed to what in ordinary life we call proper names, and, in the view I am referring to, can only be called so crudely. There is a widespread tendency to regard what in ordinary life is called a proper name only as a rough approximation of what ideally could be called so. Compare Russell's idea of the 'individual'. He talks of individuals as the ultimate constituents of reality, but says that it is difficult to say which things are individuals. The idea is that further analysis has to reveal this. We, on the other hand, introduced the idea of a proper name in a language in which it was applied to what in ordinary life we call "objects", "things" ["building stones"].

——— "What does the word 'exactness' mean? Is it real exactness if you are supposed to come to tea at 4.30 and come when a good clock strikes 4.30? Or would it only be exactness if you began to open the door at the moment the clock began to strike? But how is this moment to be defined and how is 'beginning to open the door' to be defined? Would it be correct to say, 'It is difficult to say what real exactness is, for all we know is only rough approximations'?")

5).   Questions and answers: A asks, "How many slabs?" B counts them and answers with the numeral.

Systems of communication as for instance 1), 2), 3), 4), 5) we shall call "language games". They are more or less akin to what in ordinary language we call games. Children are taught their native language by means of such games, and here they even have the entertaining character of games. We are not, however, regarding the language games which we describe as incomplete parts of a language, but as languages complete in themselves, as complete systems of human communication. To keep this point of view in mind, it very often is useful to imagine such a simple language to be the entire system of communication of a tribe in a primitive state of society. Think of primitive arithmetics of such tribes.

When the boy or grown-up learns what one might call special technical languages, e.g., the use of charts and diagrams, descriptive geometry, chemical symbolism, etc., he learns more language games. (Remark: The picture we have of the language of the grown-up is that of a nebulous mass of language, his mother tongue, surrounded by discrete and more or less clear-cut language games, the technical languages.)

6). Asking for the name: we introduce new forms of building stones. B points to one of them and asks, "What is this?"; A answers, "This is a . . .". Later on A calls out this new word, say "arch", and B brings the stone. The words, "This is . . ." together with the pointing gesture we shall call ostensive explanation or ostensive definition. In case 6) a generic name was explained, in actual fact, as the name of a shape. But we can ask analogously for the proper name of a particular object, for the name of a colour, of a numeral, of a direction.

(Remark: Our use of expressions like "names of numbers", "names of colours", "names of materials", "names of nations" may spring from two different sources. One is that we might imagine the functions of proper names, numerals, words for colours, etc., to be much more alike than they actually are. If we do so we are tempted to think that the function of every word is more or less like the function of a proper name of a person, or such generic names as "table", "chair", "door", etc. The second source is this, that if we see how fundamentally different the functions of such words as "table", "chair", etc., are from those of proper names, and how different from either the functions of, say, the names of colours, we see no reason why we shouldn't speak of names of numbers or names of directions either, not by way of saying some such thing as "numbers and directions are just different forms of objects", but rather by way of stressing the analogy which lies in the lack of analogy between the functions of the words "chair" and "Jack" on the one hand, and "east" and "Jack" on the other hand.)

7). B has a table in which written signs are placed opposite to pictures of objects (say, a table, a chair, a tea-cup, etc.). A writes one of the signs, B looks for it in the table, looks or points with his finger from the written sign to the picture opposite, and fetches the object which the picture represents.

Let us now look at the different kinds of signs which we have introduced. First let us distinguish between sentences and words. A sentence* I will call every complete sign in a language game, its constituent signs are words. (This is merely a rough and general remark about the way I will use the words "proposition"* and "word".) A proposition may consist of only one word. In 1) the signs "brick!", "column!" are the sentences. In 2) a sentence consists of two words. According to the role which propositions play in a language game, we distinguish between orders, questions, explanations, descriptions, and so on.

8). If in a language game similar to 1) A calls out an order: "slab, column, brick!" which is obeyed by B by bringing a slab, a

---

* Here Wittgenstein uses "sentence" and "proposition" interchangeably for the German "Satz".—*Edd.*

column and a brick, we might here talk of three propositions, or of one only. If, on the other hand,

9). the order of words shows B the order in which to bring the building stones, we shall say that A calls out a proposition consisting of three words. If the command in this case took the form, "Slab, then column, then brick!" we should say that it consisted of four words (not of five). Amongst the words we see groups of words with similar functions. We can easily see a similarity in the use of the words "one", "two", "three", etc. and again one in the use of "slab", "column" and "brick", etc., and thus we distinguish parts of speech. In 8) all the words of the proposition belonged to the same part of speech.

10). The order in which B had to bring the stones in 9) could have been indicated by the use of the ordinals thus: "Second, column; first, slab; third, brick!". Here we have a case in which what was the function of the order of words in one language game is the function of particular words in another.

Reflections such as the preceding will show us the infinite variety of the functions of words in propositions, and it is curious to compare what we see in our examples with the simple and rigid rules which logicians give for the construction of propositions. If we group words together according to the similarity of their functions, thus distinguishing parts of speech, it is easy to see that many different ways of classification can be adopted. We could indeed easily imagine a reason for not classing the word "one" together with "two", "three", etc., as follows:

11). Consider this variation of our language game 2). Instead of calling out, "One slab!", "One cube!", etc., A just calls "Slab!", "Cube!", etc., the use of the other numerals being as described in 2). Suppose that a man accustomed to this form 11) of communication was introduced to the use of the word "one" as described in

2). We can easily imagine that he would refuse to classify "one" with the numerals "2", "3", etc.

(Remark: Think of the reasons for and against classifying '0' with the other cardinals. "Are black and white colours?" In which cases would you be inclined to say so and which not?—Words can in many ways be compared to chess men. Think of the several ways of distinguishing different kinds of pieces in the game of chess [e.g., pawns and 'officers'].

Remember the phrase, "two or more".)

It is natural for us to call gestures, as those employed in 4), or pictures as in 7), elements or instruments of language. (We talk sometimes of a language of gestures.) The pictures in 7) and other instruments of language which have a similar function I shall call patterns. (This explanation, as others which we have given, is vague, and meant to be vague.) We may say that words and patterns have different kinds of functions. When we make use of a pattern we compare something with it, e.g., a chair with the picture of a chair. We did not compare a slab with the word "slab". In introducing the distinction, 'word/pattern', the idea was not to set up a final logical duality. We have only singled out two characteristic kinds of instruments from the variety of instruments in our language. We shall call "one", "two", "three", etc., words. If instead of these signs we used "—", "— —", "— — —", "— — — —", we might call these patterns. Suppose in a language the numerals were "one", "one one", "one one one", etc., should we call "one" a word or a pattern? The same element may in one place be used as word and in another as pattern. A circle might be the name for an ellipse, or on the other hand a pattern with which the ellipse is to be compared by a particular method of projection. Consider also these two systems of expression:

12). A gives B an order consisting of two written symbols, the first an irregularly shaped patch of a certain colour, say green, the second the drawn outline of a geometrical figure, say a circle. B brings an object of this outline and that colour, say a circular green object.

13). A gives B an order consisting of one symbol, a geometrical figure painted a particular colour, say a green circle. B brings him a green circular object. In 12) patterns correspond to our names of colours and other patterns to our names of shape. The symbols in 13) cannot be regarded as combinations of two such elements. A word in inverted commas can be called a pattern. Thus in the sentence "He said 'Go to hell'", "Go to hell" is a pattern of what he said. Compare these cases: a) Someone says "I whistled . . ." (whistling a tune); b) Someone writes, "I whistled ". An onomatopoeic word like "rustling" may be called a pattern. We call a very great variety of processes "comparing an object with a pattern". We comprise many kinds of symbols under the name "pattern". In 7) B compares a picture in the table with the objects he has before him. But what does comparing a picture with the object consist in? Suppose the table showed: a) a picture of a hammer, of pincers, of a saw, of a chisel; b) on the other hand, pictures of twenty different kinds of butterflies. Imagine what the comparison in these two cases would consist in, and note the difference. Compare with these cases a third case c) where the pictures in the table represent building stones drawn to scale, and the comparing has to be done with ruler and compasses. Suppose that B's task is to bring a piece of cloth of the colour of the sample. How are the colours of sample and cloth to be compared? Imagine a series of different cases:

14). A shows the sample to B, upon which B goes and fetches the material 'from memory'.

15). A gives B the sample, B looks from the sample to the materials on the shelves from which he has to choose.

16). B lays the sample on each bolt of material and chooses that one which he can't distinguish from the sample, for which the difference between the sample and the material seems to vanish.

17). Imagine on the other hand that the order has been, "Bring a material slightly darker than this sample". In 14) I said that B fetches the material 'from memory', which is using a common form of expression. But what might happen in such a case of comparing 'from memory' is of the greatest variety. Imagine a few instances:

14a). B has a memory image before his mind's eye when he goes for the material. He alternately looks at materials and recalls his image. He goes through this process with, say, five of the bolts, in some instances saying to himself, "Too dark", in some instances saying to himself, "Too light". At the fifth bolt he stops, says, "That's it" and takes it from the shelf.

14b). No memory image is before B's eye. He looks at four bolts, shaking his head each time, feeling some sort of mental tension. On reaching the fifth bolt, this tension relaxes, he nods his head, and takes the bolt down.

14c). B goes to the shelf without a memory image, looks at five bolts one after the other, takes the fifth bolt from the shelf.

'But this can't be all comparing consists in.'

When we call these three preceding cases, cases of comparing from memory, we feel that their description is in a sense unsatisfactory, or incomplete. We are inclined to say that the description has left out the essential feature of such a process and given us accessory features only. The essential feature it seems would be what one might call a specific experience of comparing and of recognizing. Now it is queer that on closely looking at cases of comparing, it is very easy to see a great number of activities and states of mind, all *more or less* characteristic of the act of comparing. This in fact is so, whether we speak of comparing from memory or of comparing by means of a sample before our eyes. We know a *vast* number of such processes, processes similar to each other in a vast number of different ways. We hold pieces whose colours we want to compare together or near each other for a longer or shorter period, look at them alternately or simultaneously, place them under different lights, say different things while we do so, have memory images, feelings of tension and relaxation, satisfaction and dissatisfaction, the various feelings of strain in and around our eyes accompanying prolonged gazing at the same object, and all possible combinations of these and many other experiences. The more such cases we observe and the closer we look at them, the more doubtful we feel about finding one particular mental experience characteristic of comparing. In fact, if after you had scrutinized a number of such *closely*, I admitted that there existed a peculiar mental experience which you might call the experience of comparing, and that if you insisted, I should be willing to adopt the word "comparing" only for cases in which this peculiar feeling had occurred, you would now feel that the assumption of such a peculiar experience had lost its point, because this experience was placed side by side with a vast number of other experiences which after we have scrutinized the cases seems to be that which really constitutes what connects all the cases of comparing. For the 'specific experience' we had been looking for was meant to have played the role which has been assumed by the mass of experiences revealed to us by our scrutiny: We never wanted the specific experience to be just one among a number of *more or less* characteristic experiences. (One might say that there are two ways of looking at this matter, one as it were, at close quarters, the other as though from a distance and through the medium of a peculiar atmosphere.) In fact we have found that the use which we really make of the word "comparing" is different from that which looking at it from far away we were led to expect. We find that what connects all the cases of

comparing is a vast number of overlapping similarities, and as
soon as we see this, we feel no longer compelled to say that there
must be some one feature common to them all. What ties the
ship to the wharf is a rope, and the rope consists of fibres, but it
does not get its strength from any fibre which runs through it
from one end to the other, but from the fact that there is a vast
number of fibres overlapping.

'But surely in case 14c) B acted entirely automatically. If all
that happened was really what was described there, he did not
know why he chose the bolt he did choose. He had no reason for
choosing it. If he chose the right one, he did it as a machine
might have done it'. Our first answer is that we did not deny that
B in case 14c) had what we should call a personal experience, for
we did not say that he didn't see the materials from which he
chose or that which he chose, nor that he didn't have muscular
and tactile sensations and suchlike while he did it. Now what
would such a reason which justified his choice and made it
non-automatic be like? (i.e.: What do we *imagine* it to be like?) I
suppose we should say that the opposite of automatic comparing,
as it were, the ideal case of conscious comparing, was that of having a clear memory image before our mind's eye or of seeing a
real sample and of having a specific feeling of not being able to
distinguish in a particular way between these samples and the
material chosen. I suppose that this peculiar sensation is the reason, the justification, for the choice. This specific feeling, one
might say, connects the two experiences of seeing the sample, on
the one hand, and the material on the other. But if so, what connects this specific experience with either? We don't deny that
such an experience might intervene. But looking at it as we did
just now, the distinction between automatic and non-automatic
appears no longer clear-cut and final as it did at first. We don't
mean that this distinction loses its practical value in particular
cases, e.g., if asked under particular circumstances "Did you take
this bolt from the shelf automatically, or did you think about it?",

we may be justified in saying that we did not act automatically and give as an explanation that we had looked at the material carefully, had tried to recall the memory image of the pattern, and had uttered to ourselves doubts and decisions. This may *in the particular case* be taken to distinguish automatic from non-automatic. In another case, however, we may distinguish between an automatic and a non-automatic way of the appearance of a memory image, and so on.

If our case 14c) troubles you, you may be inclined to say: "But *why* did he bring just this bolt of material? How has he recognized it as the right one? What by?"—If you ask 'why', do you ask for the cause or for the reason? If for the cause, it is easy enough to think up a physiological or psychological hypothesis which explains this choice under the given conditions. It is the task of the experimental sciences to test such hypotheses. If on the other hand you ask for a reason the answer is, "There need not have been a reason for the choice. A reason is a step preceding the step of the choice. But why should every step be preceded by another one?"

'But then B didn't really *recognize* the material as the right one.'—You needn't reckon 14c) among the cases of recognizing, but if you have become aware of the fact that the processes which we call processes of recognition form a vast family with overlapping similarities, you will probably feel not disinclined to include 14c) in this family, too.—'But doesn't B in this case lack the criterion by which he can recognize the material? In 14a), e.g., he had the memory image and he recognized the material he looked for by its agreement with the image.'—But had he also a picture of this agreement before him, a picture with which he could compare the agreement between the pattern and the bolt to see whether it was the right one? And, on the other hand, couldn't he have been given such a picture? Suppose, e.g., that A wished B to remember that what was wanted was a bolt exactly like the sample, not, as perhaps in other cases, a material slightly darker than the pattern.

Couldn't A in this case have given to B an example of the agreement required by giving him two pieces of the same colour (e.g., as a kind of reminder)? Is any such link between the order and its execution necessarily the last one?—And if you say that in 14b) at least he had the relaxing of the tension by which to recognize the right material, had he to have an image of this relaxation about him to recognize it as that by which the right material was to be recognized?—

'But supposing B brings the bolt, as in 14c), and on comparing it with the pattern it turns out to be the wrong one?'—But couldn't that have happened in all the other cases as well? Suppose in 14a) the bolt which B brought back was found not to match with the pattern. Wouldn't we in some such cases say that his memory image had changed, in others that the pattern or the material had changed, in others again that the light had changed? It is not difficult to invent cases, imagine circumstances, in which each of these judgments would be made.—'But isn't there after all an essential difference between the cases 14a) and 14c)?'— Certainly! Just that pointed out in the description of these cases.—

In 1) B learnt to bring a building stone on hearing the word "column!" called out. We could imagine what happened in such a case to be this: In B's mind the word called out brought up an image of a column, say; the training had, as we should say, established this association. B takes up that building stone which conforms to his image.—But was this *necessarily* what happened? If the training could bring it about that the idea or image— automatically—arose in B's mind, why shouldn't it bring about B's *actions* without the intervention of an image? This would only come to a slight variation of the associative mechanism. Bear in mind that the image which is brought up by the word is not arrived at by a rational process (but if it is, this only pushes our argument further back), but that this case is strictly comparable with that of a mechanism in which a button is pressed and an indicator plate appears. In fact this sort of mechanism can be used instead of that of association.

Mental images of colours, shapes, sounds, etc., etc., which play a role in communication by means of language we put in the same category with patches of colour actually seen, sounds heard.

18). The object of the training in the use of tables (as in 7) may be not only to teach the use of one particular table, but it may be to enable the pupil to use or construct himself tables with new co-ordinations of written signs and pictures. Suppose the first table a person was trained to use contained the four words "hammer", "pincers", "saw", "chisel" and the corresponding pictures. We might now add the picture of another object which the pupil had before him, say of a plane, and correlate with it the word "plane". We shall make the correlation between this new picture and word as similar as possible to the correlations in the previous table. Thus we might add the new word and picture on the same sheet, and place the new word under the previous words and the new picture under the previous pictures. The pupil will now be encouraged to make use of the new picture and word without the special training which we gave him when we taught him to use the first table. These acts of encouragement will be of various kinds, and many such acts will only be possible if the pupil responds, and responds in a particular way. Imagine the gestures, sounds, etc., of encouragement you use when you teach a dog to retrieve. Imagine on the other hand, that you tried to teach a cat to retrieve. As the cat will not respond to your encouragement, most of the acts of encouragement which you performed when you trained the dog are here out of the question.

19). The pupil could also be trained to give things names of his own invention and to bring the objects when the names are called. He is, e.g., presented with a table on which he finds pictures of objects around him on one side and blank spaces on the other, and he plays the game by writing signs of his own invention opposite the pictures and reacting in the previous way when these signs are used as orders. Or else

20). the game may consist in B's constructing a table and obeying orders given in terms of this table. When the use of a table is taught, and the table consists, say, of two vertical columns, the left hand one containing the names, the right hand one the pictures, a name and a picture being correlated by standing on a horizontal line, an important feature of the training may be that which makes the pupil slide his finger from left to right, as it were the training to draw a series of horizontal lines, one below the other. Such training may help to make the transition from the first table to the new item.

Tables, ostensive definitions, and similar instruments I shall call rules, in accordance with ordinary usage. The use of a rule can be explained by a further rule.

21). Consider this example: We introduce different ways of reading tables. Each table consists of two columns of words and pictures, as above. In some cases they are to be read horizontally from left to right, i.e., according to the scheme:

In others according to such schemes as:

or:

etc.

Schemes of this kind can be adjoined to our tables, as rules for reading them. Could not these rules again be explained by further rules? Certainly. On the other hand, is a rule incompletely explained if no rule for its usage has been given?

We introduce into our language games the endless series of numerals. But how is this done? Obviously the analogy between this process and that of introducing a series of twenty numerals is not the same as that between introducing a series of twenty numerals and introducing a series of ten numerals. Suppose that our game was like 2) but played with the endless series of numerals. The difference between it and 2) would not be just that more numerals were used. That is to say, suppose that as a matter of fact in playing the game we had actually made use of, say, 155 numerals, the game we play would not be that which could be described by saying that we played the game 2), only with 155 instead of 10 numerals. But what does the difference consist in? (The difference would seem to be almost one of the spirit in which the games are played.) The difference between games can lie, say, in the number of the counters used, in the number of squares of the playing board, or in the fact that we use squares in one case and hexagons in the other, and suchlike. Now the difference between the finite and infinite game does not seem to lie in the material tools of the game; for we should be inclined to say that infinity can't be expressed in them, that is, that we can only conceive of it in our thoughts, and hence that it is in these thoughts that the finite and infinite game must be distinguished. (It is queer though that these thoughts should be capable of being expressed in signs.)

Let us consider two games. They are both played with cards carrying numbers, and the highest number takes the trick.

22). One game is played with a fixed number of such cards, say 32. In the other game we are under certain circumstances allowed to increase the number of cards to as many as we like, by cutting pieces of paper and writing numbers on them. We will call the first of these games bounded, the second unbounded. Suppose a hand of the second game was played and the number of cards actually used was 32. What is the difference in this case between playing a hand *a*) of the unbounded game and playing a hand *b*) of the bounded game?

The difference will not be that between a hand of a bounded game with 32 cards and a hand of a bounded game with a greater number of cards. The number of cards used was, we said, the same. But there will be differences of another kind, e.g., the bounded game is played with a normal pack of cards, the unbounded game with a large supply of blank cards and pencils. The unbounded game is opened with the question, "How high shall we go?" If the players look up the rules of this game in a book of rules, they will find the phrase "and so on" or "and so on *ad inf.*" at the end of certain series of rules. So the difference between the two hands *a*) and *b*) lies in the tools we use, though admittedly not in the cards they are played with. But this difference seems trivial and not the essential difference between the games. We feel that there must be a big and essential difference somewhere. But if you look closely at what happens when the hands are played, you find that you can only detect a number of differences in details, each of which would seem inessential. The acts, e.g., of dealing and playing the cards *may* in both cases be identical. In the course of playing the hand *a*), the players may have considered making up more cards, and again discarded the idea. But what was it like to consider this? It could be some such process as saying to themselves or aloud "I wonder whether I should make up another card". Again, no such consideration may have entered the minds of the players. It is possible that the whole difference in the events of a hand of the bounded, and a hand of the unbounded, game lay in what was said before the game started, e.g., "Let's play the bounded game".

'But isn't it correct to say that hands of the two different games belong to two different systems?' Certainly. Only the facts which we are referring to by saying that they belong to different systems are much more complex than we might expect them to be.

Let us now compare language games of which we should say that they are played with a limited set of numerals, with language games of which we should say that they are played with the endless series of numerals.

23). Like 2) A orders B to bring him a number of building stones. The numerals are the signs "1", "2", ... "9", each written on a card. A has a set of these cards and gives B the order by showing him one of the set and calling out one of the words, "slab", "column", etc.

24). Like 23), only there is no set of indexed cards. The series of numerals 1 . . . 9 is learned by heart. The numerals are called out in the orders, and the child learns them by word of mouth.

25). An abacus is used. A sets the abacus, gives it to B, B goes with it to where the slabs lie, etc.

26). B is to count the slabs in a heap. He does it with an abacus, the abacus has twenty beads. There are never more than 20 slabs in a heap. B sets the abacus for the heap in question and shows A the abacus thus set.

27). Like 26). The abacus has 20 small beads and one large one. If the heap contains more than 20 slabs, the large bead is moved. (So the large bead in some way corresponds to the word "many".)

28). Like 26). If the heap contains n slabs, n being more than 20 but less than 40, B moves n-20 beads, shows A the abacus thus set and claps his hands once.

29). A and B use the numerals of the decimal system (written or spoken) up to 20. The child learning this language learns these numerals by heart, as in 2).

30). A certain tribe has a language of the kind 2). The numerals used are those of our decimal system. No one numeral used can be observed to play the predominant role of the last numeral in some of the above games [27), 28)]. (One is tempted to continue this sentence by saying, "although there is of course a highest numeral actually used".) The children of the tribe learn the numerals in this way: They are taught the signs from 1 to 20 as in 2) and to count rows of beads of no more than 20 on being ordered, "Count these". When in counting the pupil arrives at the numeral 20, one makes a gesture suggestive of "Go on", upon which the child says (in most cases at any rate) "21". Analogously, the children are made to count to 22 and to higher numbers, no particular number playing in these exercises the predominant role of a last one. The last stage of the training is that the child is ordered to count a group of objects, well above 20, without the suggestive gesture being used to help the child over the numeral 20. If a child does not respond to the suggestive gesture, it is separated from the others and treated as a lunatic.

31). Another tribe. Its language is like that in 30). The highest numeral observed in use is 159. In the life of this tribe the numeral 159 plays a peculiar role. Supposing I said, "They treat this number as their highest",——but what does this mean? Could we answer: "They just say that it is the highest"?—They say certain words, but how do we know what they mean by them? A criterion for what they mean would be the occasions on which the word we are inclined to translate into our word "highest" is used, the role, we might say, which we observe this word to play in the life of the tribe. In fact we could easily imagine the numeral 159 to be used on such occasions, in connection with such gestures and forms of behaviour as would make us say that this numeral plays the role of an unsurmountable limit, even if the tribe had no word corresponding to our "highest", and the criteria for numeral 159 being the highest numeral did not consist of anything that was *said* about the numeral.

32). A tribe has two systems of counting. People learned to count with the alphabet from A to Z and also with the decimal system as in 30.) If a man is to count objects with the first system, he is ordered to count "*in the closed way*", in the second case, "*in the open way*"; and the tribe uses the words "closed" and "open" also for a closed and open door.

(Remarks: 23) is limited in an obvious way by the set of cards. 24): Note analogy and lack of analogy between the *limited supply* of cards in 23) and of words in our memory in 24). Observe that the limitation in 26) on the one hand lies in the *tool* [the abacus of 20 beads] and its usage in our game, on the other hand [in a totally different way] in the fact that in the actual practice of playing the game no more than 20 objects are ever to be counted. In 27) that latter kind of limitation was absent, but the large bead rather stressed the limitation of our means. Is 28) a limited or an unlimited game? The practice we have described gives the limit 40. We are inclined to say this game 'has it in it' to be continued indefinitely, but remember that we could also have construed the preceding games as beginnings of a system. In 29) the systematic aspect of the numerals used is even more conspicuous than in 28). One might say that there was no limitation imposed by the tools of this game, if it were not for the remark that the numerals up to 20 are learnt by heart. This suggests the idea that the child is not taught to '*understand*' the system which we see in the decimal notation. Of the tribe in 30) we should certainly say that they are trained to construct numerals indefinitely, that the arithmetic of their language is not a finite one, that their series of numbers has no end. [It is just in such a case when numerals are constructed 'indefinitely' that we say that people have the infinite series of numbers.] 31) might show you what a vast variety of cases can be imagined in which we should be inclined to say that the arithmetic of the tribe deals with a finite series of numbers, even in spite of the fact that the way in which the children are trained in the use of numerals suggests no upper limit. In 32) the terms "closed" and "open" [which could by a slight variation of the example be replaced by "limited" and "unlimited"] are introduced into the language of the tribe itself. Introduced in that simple and clearly circumscribed game, there is of course nothing mysterious about the use of the word "open". But this word corresponds to our "infinite", and the games we play with the latter differ from 31) only by being vastly more complicated. In other words, our use of the word "infinite" is just as *straightforward* as that of "open" in 31), and our idea that its meaning is 'transcendent' rests on a misunderstanding.)

We might say roughly that the unlimited cases are characterized by this: that they are not played with a *definite supply* of numerals, but instead with a *system* for constructing numerals (indefinitely). When we say that someone has been supplied with a system for constructing numerals, we generally think of one of three things: a) of giving him a *training* similar to that described in 30), which, experience teaches us, will make him pass tests of the kind mentioned there; b) of creating a *disposition* in the same man's mind, or brain, to react in that way; c) of supplying him with a *general* rule for the construction of numerals.

What do we call a rule? Consider this example:

33). B moves about according to rules which A gives him.

B is supplied with the following table:

A gives an order made up of the letters in the table, say: "aa-caddd". B looks up the arrow corresponding to each letter of the order and moves accordingly; in our example thus: →→→↑→↓↓↓

The table 33) we should call a rule (or else "the expression of a rule". Why I give these synonymous expressions will appear later.) We shan't be inclined to call the sentence "aacaddd" itself a rule. It is of course the description of the way B has to take. On the other hand, such a description would under certain circumstances be called a rule, e.g., in the following case:

34). B is to draw various ornamental linear designs. Each design is a repetition of one element which A gives him. Thus if A gives the order "cada", B draws a line thus: ⌐_⌐_⌐_⌐_

In this case I think we should say that "cada" is the rule for drawing the design. Roughly speaking, it characterizes what we call a rule to be applied repeatedly, in an indefinite number of instances. Cf., e.g., the following case with 34):

35). A game played with pieces of various shapes on a chess board. The way each piece is allowed to move is laid down by a rule. Thus the rule for a particular piece is "ac", for another piece "acaa", and so on. The first piece then can make a move like this: →↑, the second, like this: →↑→→. Both a formula like "ac" or a diagram like that corresponding to such a formula might here be called a rule.

36). Suppose that after playing the game 33) several times as described above, it was played with this variation: that B no longer looked at the table, but reading A's order the letters call up the images of the arrows (by association), and B acts according to these imagined arrows.

37). After playing it like this for several times, B moves about according to the written order as he would have done had he looked up or imagined the arrows, but actually without any such picture intervening. Imagine even this variation:

38). B in being trained to follow a written order, is shown the table of 33) once, upon which he obeys A's orders without further intervention of the table in the same way in which B in 33) does with the help of the table on each occasion.

In each of these cases, we might say that the table 33) is a rule of the game. But in each one this rule plays a different role. In 33) the table is an instrument used in what we should call *the practice* of the game. It is replaced in 36) by the working of association. In 37) even this shadow of the table has dropped out of the practice of the game, and in 38) the table is admittedly an instrument for the *training* of B only.

But imagine this further case:

39). A certain system of communication is used by a tribe. I will describe it by saying that it is similar to our game 38) except that no table is used in the training. The training *might* have consisted in several times leading the pupil by the hand along the path one wanted him to go. But we could also imagine a case:

40). where even this training is not necessary, where, as we should say, the look of the letters abcd naturally produced an urge to move in the way described. This case at first sight looks puzzling. We seem to be assuming a most unusual working of the mind. Or we may ask, "How on earth is he to know which way to move if the letter a is shown him?". But isn't B's reaction in this case the very reaction described in 37) and 38), and in fact our usual reaction when for instance we hear and obey an order? For, the fact that the training in 38) and 39) *preceded* the carrying out of the order does not change the process of carrying out. In other words the 'curious mental mechanism' assumed in 40) is no other than that which we assumed to be created by the training in 37) and 38). 'But *could* such a mechanism be born with you?' But did you find any difficulty in assuming that *that* mechanism was born with B, which enabled him to respond to the training in the way he did? And remember that the rule or explanation given in table 33) of the signs abcd was not essentially the last one, and that we might have given a table for the use of such tables, and so on. (Cf. 21).)

How does one explain to a man how he should carry out the order, "Go *this* way!" (pointing with an arrow the way he should go)? Couldn't this mean going the direction which we should call the opposite of that of the arrow? Isn't every explanation of how he should follow the arrow in the position of another arrow? What would you say to this explanation: A man says, "If I point this way (pointing with his right hand) I mean you to go like this" (pointing with his left hand the same way)? This just shows you the extremes between which the uses of signs vary.

Let us return to 39). Someone visits the tribe and observes the use of the signs in their language. He describes the language by saying that its sentences consist of the letters abcd used according to the table [of 33)]. We see that the expression, "A game is played according to the rule so and so" is used not only in the variety of cases exemplified by 36), 37), and 38), but even in cases where the rule is neither an instrument of the training nor of the practice of the game, but stands in the relation to it in which our table stands to the practice of our game 39). One might in this case call the table a natural law describing the behaviour of the people of this tribe. Or we might say that the table is a record belonging to the natural history of the tribe.

Note that in the game 33) I distinguished sharply between the order to be carried out and the rule employed. In 34), on the other hand, we called the sentence "cada" a rule, and it was the order. Imagine also this variation:

41). The game is similar to 33), but the pupil is not just trained to use a single table; but the training aims at making the pupil use any table correlating letters with arrows. Now by this I mean no more than that the training is of a peculiar kind, roughly speaking one analogous to that described in 30). I will refer to a training more or less similar to that in 30) as a *"general training"*. General trainings form a family whose members differ greatly from one another. The kind of thing I'm thinking of now mainly consists: *a*) of a training in a limited range of actions, *b*) of giving the pupil a lead to extend this range, and *c*) of random exercises and tests. After the general training the order is now to consist in giving him a sign of this kind:

He carries out the order by moving thus:

Here I suppose we should say the table, the rule, is *part* of the order.

Note, we are not saying '*what a rule is*' but just giving different applications of the word "rule": and we certainly do this by giving applications of the words "expression of a rule".

Note also that in 41) there is no clear case against calling the whole symbol given the sentence, though we *might* distinguish in it between the sentence and the table. What in this case more particularly tempts us to this distinction is the linear writing of the part outside the table. Though from certain points of view we should call the linear character of the sentence merely external and inessential, this character and similar ones play a great role in what as logicians we are inclined to say about sentences and propositions. And therefore if we conceive of the symbol in 41) as a unit, this may make us realize what a sentence *can* look like.

Let us now consider these two games:

42). A gives orders to B: They are written signs consisting of dots and dashes, and B executes them by doing a figure in dancing with a particular step. Thus the order "—." is to be carried out by taking a step and a hop alternately; the order ". . —— —— —" by alternately taking two hops and three steps, etc. The training in this game is 'general' in the sense explained in 41); and I should like to say, "The orders given don't move in a limited range. They comprise combinations of any number of dots and dashes".—But what does it mean to say that the orders don't move in a limited range? Isn't this nonsense? Whatever orders are given in the practice of the game constitute the limited range.— Well, what I meant to say by "The orders don't move in a limited range" was that neither in the teaching of the game nor in the practice of it a limitation of the range plays a 'predominant' role (see 30)), or, as we might say, the range of the game (it is superfluous to say limited) is just the extent of its actual ('accidental') practice. (Our game is in this way like 30).) Cf. with this game the following:

43). The orders and their execution as in 42); but only these three signs are used: "—", "— · ·", ". — —". We say that in 42) B, in executing the order, is *guided* by the sign given to him. But if we ask ourselves whether the three signs in 43) guide B in executing the orders, it seems that we can say both yes and no according to the way we look at the execution of the orders.

If we try to decide whether B in 43) is guided by the signs or not, we are inclined to give such answers as the following: *a*) B is guided if he doesn't just look at an order, say ". —— —" as a whole and then act, but if he reads it 'word by word' (the words used in our language being "." and "—") and acts according to the words he has read.

We could make these cases clearer if we imagine that the 'reading word by word' consisted in pointing to each word of the sentence in turn with one's finger as opposed to pointing at the whole sentence at once, say by pointing to the beginning of the sentence. And the 'acting according to the words' we shall for the sake of simplicity imagine to consist in acting (stepping or hopping) after each word of the sentence in turn.— *b*) B is guided if he goes through a conscious process which makes a connection between the pointing to a word and the act of hopping and stepping. Such a connection could be imagined in many different ways. E.g., B has a table in which a dash is correlated to the picture of a man making a step and a dot to a picture of a man hopping. Then the conscious acts connecting reading the order and carrying it out might consist in consulting the table, or in consulting a memory image of it 'with one's mine's eye'. *c*) B is guided if he does not just react to looking at each word of the order, but experiences the peculiar strain of 'trying to remember what the sign means', and further, the relaxing of this strain when the meaning, the right action, comes before his mind.

All these explanations seem in a peculiar way unsatisfactory, and it is the limitation of our game which makes them unsatisfactory. This is expressed by the explanation that B is guided by the particular combination of words in one of our three sentences if he *could* also have carried out orders consisting in other combinations of dots and dashes. And if we say this, it seems to us that the '*ability*' to carry out other orders is a particular state of the person carrying out the orders of 42). And at the same time we can't in this case find anything which we should call such a state.

Let us see what role the words "can" or "to be able to" play in our language. Consider these examples:

44). Imagine that for some purpose or other people use a kind of instrument or tool; this consists of a board with a slot in it guiding the movement of a peg. The man using the tool slides the peg along the slot. There are such boards with straight slots, circular slots, elliptic slots, etc. The language of the people using this instrument has expressions for describing the activity of moving the peg in the slot. They talk of moving it in a circle, in a straight line, etc. They also have a means of describing the board used. They do it in this form: "This is a board in which the peg *can* be moved in a circle". One could in this case call the word "can" an operator by means of which the form of expression describing an action is transformed into a description of the instrument.

45). Imagine a people in whose language there is no such form of sentence as "the book is in the drawer" or "water is in the glass", but wherever we should use these forms they say, "The book can be taken out of the drawer", "The water can be taken out of the glass".

46). An activity of the men of a certain tribe is to test sticks as to their hardness. They do it by trying to bend the sticks with their hands. In their language they have expressions of the form, "This stick can be bent easily", or "This stick can be bent with difficulty". They use these expressions as we use "This stick is soft", or "This stick is hard". I mean to say that they don't use the expression, "This stick can be bent easily" as we should use the sentence, "I am bending the stick with ease". Rather they use their expression in a way which would make us say that they are describing a state of the stick. I.e., they use such sentences as, "This hut is built of sticks that can be bent easily". (Think of the way in which we form adjectives out of verbs by means of the ending "able", e.g., "deformable".)

Now we might say that in the last three cases the sentences of the form "so and so can happen" described the state of objects, but there are great differences between these examples. In 44) we saw the state described before our eyes. We saw that the board had a circular or a straight slot, etc. In 45), in some instances at least this was the case, we could see the objects in the box, the water in the glass, etc. In such cases we use the expression "state of an object" in such a way that there corresponds to it what one might call a stationary sense experience.

When, on the other hand, we talk of the state of a stick in 46), observe that to this 'state' there does not correspond a particular sense experience which lasts while the state lasts. Instead of that, the defining criterion for something being in this state consists in certain *tests*.

We may say that a car travels 20 miles an hour even if it only travels for half an hour. We can explain our form of expression by saying that the car travels with a speed which enables it to make 20 miles an hour. And here also we are inclined to talk of the velocity of the car as of a state of its motion. I think we should not use this expression if we had no other 'experiences of motion' than those of a body being in a particular place at a certain time and in another place at another time; if, e.g., our experiences of motion were of the kind which we have when we see that the hour hand of the clock has moved from one point of the dial to the other.

47). A tribe has in its language commands for the execution of certain actions of men in warfare, something like "Shoot!", "Run!", "Crawl!", etc. They also have a way of describing a man's build. Such a description has the form "He can run fast", "He can throw the spear far". What justifies me in saying that these sentences are descriptions of the man's build is the use which they make of sentences of this form. Thus if they see a man with bulging leg muscles but who as we should say has not the use of his legs for some reason or other, they say he is a man who can run fast. The drawn image of a man which shows large biceps they describe as representing a man "who can throw a spear far".

48). The men of a tribe are subjected to a kind of medical examination before going into war. The examiner puts the men through a set of standardized tests. He lets them lift certain weights, swing their arms, skip, etc. The examiner then gives his verdict in the form "So and so can throw a spear" or "can throw a boomerang" or "is fit to pursue the enemy", etc. There are no special expressions in the language of this tribe for the activities performed in the tests; but these are referred to only as the tests for certain activities in warfare.

It is an important remark concerning this example and others which we give that one may object to the description which we give of the language of a tribe, that in the specimens we give of their language we let* them speak English, thereby already pre-supposing the whole background of the English language, that is, our usual meanings of the words. Thus if I say that in a certain language there is no special verb for "skipping", but that this language uses instead the form "making the test for throwing the boomerang", one may ask how I have characterized the use of the expressions, "make a test for" and "throwing the boomerang", to be justified in substituting these English expressions for whatever their actual words may be. To this we must answer that we have only given a very sketchy description of the practices of our ficti-tious languages, in some cases only hints, but that one can easily make these descriptions more complete. Thus in 48) I could have said that the examiner uses orders for making the men go through the tests. These orders all begin with one particular expression

---

* German *lassen*, i.e., 'make'.—*Edd.*

which I could translate into the English words, "Go through the test". And this expression is followed by one which in actual warfare is used for certain actions. Thus there is a command upon which men throw their boomerangs and which therefore I should translate into, "Throw the boomerangs". Further, if a man gives an account of the battle to his chief, he again uses the expression I have translated into "throw a boomerang", this time in a description. Now what characterizes an order as such, or a description as such, or a question as such, etc., is—as we have said—the role which the utterance of these signs plays in the whole practice of the language. That is to say, whether a word of the language of our tribe is rightly translated into a word of the English language depends upon the role this word plays in the whole life of the tribe; the occasions on which it is used, the expressions of emotion by which it is generally accompanied, the ideas which it generally awakens or which prompt its saying, etc., etc. As an exercise ask yourself: in which cases would you say that a certain word uttered by the people of the tribe was a greeting? In which cases should we say it corresponded to our "Goodbye", in which to our "Hello"? In which cases would you say that a word of a foreign language corresponded to our "perhaps"?—to our expressions of doubt, trust, certainty? You will find that the justifications for calling something an expression of doubt, conviction, etc., largely, though of course not wholly, consist in descriptions of gestures, the play of facial expressions, and even the tone of voice. Remember at this point that the personal experiences of an emotion must in part be strictly localized experiences; for if I frown in anger I feel the muscular tension of the frown in my forehead, and if I weep, the sensations around my eyes are obviously part, and an important part, of what I feel. This is, I think, what William James meant when he said that a man doesn't cry because he is sad but that he is sad because he cries. The reason why this point is often not understood, is that we think of the utterance of an emotion as though it were some artificial device to let others know that we have it. Now there is no sharp line between such 'artificial devices' and what one might call the natural expressions of emotion. Cf. in this respect: *a*) weeping, *b*) raising one's voice when one is angry, *c*) writing an angry letter, *d*) ringing the bell for a servant you wish to scold.

49)   Imagine a tribe in whose language there is an expression corresponding to our "He has done so and so", and another expression corresponding to our "He can do so and so", this latter expression, however, being only used where its use is justified by the same fact which would also justify the former expression. Now what can make me say this? They have a form of communication which we should call narration of past events because of the circumstances under which it is employed. There are also circumstances under which we should ask and answer such questions as "Can so and so do this?" Such circumstances can be described, e.g., by saying that a chief picks men suitable for a certain action, say crossing a river, climbing a mountain, etc. As the defining criteria of "the chief picking men suitable for this action", I will not take what he says but only the other features of the situation. The chief under these circumstances asks a question which, as far as its practical consequences go, would have to be translated by our "Can so and so swim across this river?" This question however, is only answered affirmatively by those who actually have swum across this river. This answer is not given in the same words in which under the circumstances characterizing narration he would say that he has swum across this river, but it is given in the terms of the question asked by the chief. On the other hand, this answer is not given in cases in which we should certainly give the answer, "I can swim across this river", if, e.g., I had performed more difficult feats of swimming though not just that of swimming across this particular river.

By the way, have the two phrases "He has done so and so" and "He can do so and so" the same meaning in this language or have they different meanings? If you think about it, something will tempt you to say the one, something to say the other. This only shows that the question has here no clearly defined meaning. All I can say is: If the fact that they only say "He can . . ." if he has done . . . is your criterion for the same meaning, then the two expressions have the same meaning. If the circumstances under which an expression is used make its meaning, the meanings are different. The use which is made of the word "can"—the expression of possibility in 49)—can throw a light upon the idea that what can happen must have happened before (Nietzsche). It will also be interesting to look, in the light of our examples, on the statement that what happens can happen.

Before we go on with our consideration of the use of 'the expression of possibility', let us get clearer about that department of our language in which things are said about past and future, that is, about the use of sentences containing such expressions as "yesterday", "a year ago", "in five minutes", "before I did this", etc. Consider this example:

50). Imagine how a child might be trained in the practice of 'narration of past events'. He was first trained in asking for certain things (as it were, in giving orders. See 1)). Part of this training was the exercise of 'naming the things'. He has thus learnt to name (and ask for) a dozen of his toys. Say now that he has played with three of them (e.g., a ball, a stick, and a rattle), then they are taken away from him, and now the grown-up says such a phrase as, "He's had a ball, a stick, and a rattle". On a similar occasion he stops short in the enumeration and induces the child to complete it. On another occasion, perhaps, he only says, "He's had . . ." and leaves the child to give the whole enumeration. Now the way of 'inducing the child to go on' can be this: He stops short in his enumeration with a facial expression and a raised tone of voice which we should call one of expectancy. All then depends on whether the child will react to this 'inducement' or not. Now there is a queer misunderstanding we are most liable to fall into, which consists in regarding the 'outward means' the teacher uses to induce the child to go on as what we might call an indirect means of making himself understood to the child. We regard the case as though the child already possessed a language in which it thought and that the teacher's job is to induce it to guess his meaning in the realm of meanings before the child's mind, as though the child could in his own private language ask himself such a question as, "Does he want me to continue, or repeat what he said, or something else?" (Cf. with 30).)

51). Another example of a primitive kind of narration of past events: we live in a landscape with characteristic natural landmarks against the horizon. It is therefore easy to remember the place at which the sun rises at a particular season, or the place above which it stands when at its highest point, or the place at which it sets. We have some characteristic pictures of the sun in different positions in our landscape. Let us call this series of pictures the sun series. We have also some characteristic pictures of the activities of a child, lying in bed, getting up, dressing, lunching, etc. This set I'll call the life pictures. I imagine that the child can frequently see the position of the sun while about the day's activities. We draw the child's attention to the sun's standing in a certain place while the child is occupied in a particular way. We then let it look both at a picture representing its occupation and at a picture showing the sun in its position at that time. We can thus roughly tell the story of the child's day by laying out a row of the life pictures, and above it what I called the sun series, the two rows in the proper correlation. We shall then proceed to let the child supplement such a picture story, which we leave incomplete. And I wish to say at this point that this form of training [see 50) and 30)] is one of the big characteristic features in the use of language, or in thinking.

52). A variation of 51). There is a big clock in the nursery, for simplicity's sake imagine it with an hour hand only. The story of the child's day is narrated as above, but there is no sun series; instead we write one of the numbers of the dial against each life picture.

53). Note that there would have been a similar game in which also, as we might say, time was involved, that of just laying out a series of life pictures. We might play this game with the help of words which would correspond to our "before" and "after". In this sense we may say that 53) involves the ideas of before and after, but not the idea of a measurement of time. I needn't say that an easy step would lead us from the narrations in 51), 52), and 53) to narrations in words. Possibly someone considering such forms of narration might think that in them the real idea of time isn't yet involved at all, but only some crude substitute for it, the positions of a clock hand and suchlike. Now if a man claimed that there is an idea of five o'clock which does not bring in a clock, that the clock is only the coarse instrument indicating when it is five o'clock or that there is an idea of an hour which does not bring in an instrument for measuring the time, I will not contradict him, but I will ask him to explain to me what his use of the term "an hour" or "five o'clock" is. And if it is not that involving a clock, it is a different one; and then I will ask him why he uses the terms "five o'clock", "an hour", "a long time", "a short time", etc., in one case in connection with a clock, in the other independent of one; it will be because of certain analogies holding between the two uses, but we have now two uses of these terms, and no reason to say that one of them is less real and pure than the other. This might get clearer by considering the following example:

54). If we give a person the order "Say a number, any one which comes into your mind", he can generally comply with it at once. Suppose it were found that the numbers thus said on request increased—with every normal person—as the day went on; a man starts out with some small number every morning and reaches the highest number before falling asleep at night. Consider what could tempt one to call the reactions described "a means of measuring time" or even to say that they are the *real* milestones in the passage of time, the sun clocks, etc., being only indirect markers. (Examine the statement that the human heart is the real clock behind all the other clocks.)

Let us now consider further language games into which temporal expressions enter.

55). This arises out of 1). If an order like "slab!", "column!", etc. is called out, B is trained to carry it out immediately. We now introduce a clock into this game, an order is given, and we train the child not to carry it out until the hand of our clock reaches a point indicated before with the finger. (This might, e.g., be done in this way: You first trained the child to carry out the order immediately. You then give the order, but hold the child back, releasing it only when the hand of the clock has reached the point of the dial to which we point with our fingers.)

We could at this stage introduce such a word as "now". We have two kinds of orders in this game, the orders used in 1), and orders consisting of these, together with a gesture indicating a point of the clock dial. In order to make the distinction between these two kinds more explicit, we may affix a particular sign to the orders of the first kind and, e.g., say: "slab, now!".

It would be easy now to describe language games with such expressions as "in five minutes", "half an hour ago".

56). Let us now have the case of a *description* of the future, a forecast. One might, e.g., awaken the tension of expectation in a child by keeping his attention for a considerable time on some traffic lights changing their colour periodically. We also have a red, a green, and a yellow disc before us and alternately point to one of these discs by way of forecasting the colour which will appear next. It is easy to imagine further developments of this game.

Looking at these language games, we don't come across the ideas of the past, the future and the present in their problematic and almost mysterious aspect. What this aspect is and how it comes about that it appears can be almost characteristically exemplified if we look at the question "Where does the present go when it becomes past, and where is the past?"—Under what circumstances has this question an allurement for us? For under certain circumstances it hasn't, and we should wave it away as nonsense.

It is clear that this question most easily arises if we are preoccupied with cases in which there are things flowing by us,—as logs of wood float down a river. In such a case we can say the logs which *have passed* us are all down towards the left and the logs which *will pass* us are all up towards the right. We then use this situation as a simile for all happening in time and even embody the simile in our language, as when we say that 'the present event passes by' (a log passes by), 'the future event is to come' (a log is to come). We talk about the flow of events; but also about the flow of time—the river on which the logs travel.

Here is one of the most fertile sources of philosophic puzzlement: we talk of the future event of something coming into my room, and also of the future coming of this event.

We say, "*Something* will happen", and also, "Something comes towards me"; we refer to the log as "something", but also the log's coming towards me.

Thus it can come about that we aren't able to rid ourselves of the implications of our symbolism, which seems to admit of a question like "Where does the flame of a candle go to when it's blown out?" "Where does the light go to?", "Where does the past go to"? We have become obsessed with our symbolism.—We may say that we are led into puzzlement by an analogy which irresistibly drags us on.—And this also happens when the meaning of the word "now" appears to us in a mysterious light. In our example 55) it appears that the function of "now" is in no way comparable to the function of an expression like "five o'clock", "midday", "the time when the sun sets", etc. This latter group of expressions I might call "specifications of times". But our ordinary language uses the word "now" and specifications of time in similar contexts. Thus we say

"The sun sets at six o'clock".
"The sun is setting now".

We are inclined to say that both "now" and "six o'clock" 'refer to points of time'. This use of words produces a puzzlement which one might express in the question "What is the 'now'?—for it is a moment of time and yet it can't be said to be either the 'moment at which I speak' or 'the moment at which the clock strikes', etc., etc."—Our answer is: The function of the word "now" is entirely different from that of a specification of time.—This can easily be seen if we look at the role this word really plays in our usage of language, but it is obscured when instead of looking at the *whole language game*, we only look at the contexts, the phrases of language in which the word is used. (The word "today" is not a date, but it isn't anything like it either. It doesn't differ from a date as a hammer differs from a mallet, but as a hammer differs from a nail; and surely we may say there is both a connection between a hammer and a mallet and between a hammer and a nail.)

One has been tempted to say that "now" is the name of an instant of time, and this, of course, would be like saying that "here" is the name of a place, "this" the name of a thing, and "I" the name of a man. (One could, of course, also have said "a year ago" was the name of a time, "over there" the name of a place, and "you" the name of a person.) But nothing is more unlike than the use of the word "this" and the use of a proper name—I mean *the games* played with these words, not the phrases in which they are used. For we do say "This is short" and "Jack is short"; but remember that "This is short" without the pointing gesture and without the thing we are pointing to would be meaningless.—What can be compared with a name is not the word "this" but, if you like, the symbol consisting of this word, the gesture, and the sample. We might say: Nothing is more characteristic of a proper name A than that we can use it in such a phrase as "This is A"; and it makes no sense to say "This is this" or "Now is now" or "Here is here".

The idea of a proposition saying something about what will happen in the future is even more liable to puzzle us than the idea of a proposition about the past. For comparing future events with past events, one may almost be inclined to say that though the past events do not really exist in the full light of day, they exist in an underworld into which they have passed out of the real life; whereas the future events do not even have this shadowy existence. We could, of course, imagine a realm of the unborn, future events, whence they come into reality and pass into the realm of the past; and, if we think in terms of this metaphor, we may be surprised that the future should appear less existent than the past. Remember, however, that the grammar of our temporal expressions is not symmetrical with respect to an origin corresponding with the present moment. Thus the grammar of the expressions relating to memory does not reappear 'with opposite sign' in the grammar of the future tense. This is the reason why it has been said that propositions concerning future events are not really propositions. And to say this is all right as long as it isn't meant to be more than a decision about the use of the term "proposition"; a decision which, though not agreeing with the common usage of the word "proposition", may come natural to human beings under certain circumstances. If a philosopher says that propositions about the future are not real propositions, it is because he has been struck by the asymmetry in the grammar of temporal expressions. The danger is, however, that he imagines he has made a kind of scientific statement about 'the nature of the future'.

57). A game is played in this way: A man throws a die, and before throwing he draws on a piece of paper some one of the six faces of the die. If, after having thrown, the face of the die turning up is the one he has drawn, he feels (expresses) satisfaction. If a different face turns up, he is dissatisfied. Or, let there be two partners and every time one guesses correctly what he will throw his partner pays him a penny, and if incorrectly, he pays his partner. Drawing the face of the die will under the circumstances of this game be called "making a guess" or "a conjecture".

58). In a certain tribe contests are held in running, putting the weight, etc., and the spectators stake possessions on the competitors. The pictures of all the competitors are placed in a row, and what I called the spectator's staking property on one of the competitors consists in laying this property (pieces of gold) under one of the pictures. If a man has placed his gold under the picture of the winner in the competition he gets back his stake doubled. Otherwise he loses his stake. Such a custom we should undoubtedly call betting, even if we observed it in a society whose language held no scheme for stating 'degrees of probability', 'chances' and the like. I assume that the behaviour of the spectators expresses great keenness and excitement before and after the outcome of the bet is known. I further imagine that on examining the placing of the bets I can understand 'why' they were thus placed. I mean: In a competition between two wrestlers, mostly the bigger man is the favourite; or if the smaller, I find that he has shown greater strength on previous occasions, or that the bigger had recently been ill, or had neglected his training, etc.

Now this may be so although the language of the tribe does not express reasons for the placing of the bets. That is to say, nothing in their language corresponds to our saying, e.g., "I bet on this man because he has kept fit, whereas the other has neglected his training", and suchlike. I might describe this state of affairs by saying that my observation has taught me certain *causes* for their placing their bets as they do, but that the bettors used no *reasons* for acting as they did.

The tribe may, on the other hand, have a language which comprises 'giving reasons'. Now this game of giving the reason why one acts in a particular way does not involve finding the causes of one's actions (by frequent observations of the conditions under which they arise). Let us imagine this:

59). If a man of our tribe has lost his bet and upon this is chaffed or scolded, he points out, possibly exaggerating, certain features of the man on whom he has laid his bet. One can imagine a discussion of pros and cons going on in this way: two people pointing out alternately certain features of the two competitors whose chances, as we should say, they are discussing; A pointing with a gesture to the great height of the one, B in answer to this shrugging his shoulders and pointing to the size of the other's biceps, and so on. I could easily add more details which would make us say that A and B are giving reasons for laying a bet on one person rather than on the other.

Now one might say that giving reasons in this way for laying their bets certainly presupposes that they have observed causal connections between the result of a fight, say, and certain features of the bodies of the fighters, or of their training. But this is an assumption which, whether reasonable or not, I certainly have not made in the description of our case. (Nor have I made the assumption that the bettors give reasons for their reasons.) We should in a case like that just described not be surprised if the language of the tribe contained what we should call expressions of degrees of belief, conviction, certainty. These expressions we could imagine to consist in the use of a particular word spoken with different intonations, or a series of words. (I am not thinking, however, of the use of a scale of probabilities.)—It is also easy to imagine that the people of our tribe accompany their betting by verbal expressions which we translate into "I believe that so and so *can* beat so and so in wrestling", etc.

60). Imagine in a similar way conjectures being made as to whether a certain load of gunpowder will be sufficient to blast a certain rock, and the conjecture to be expressed in a phrase of the form "This quantity of gunpowder can blast this rock".

61). Compare with 60) the case in which the expression "I shall be able to lift this weight" is used as an abbreviation for the conjecture "My hand holding this weight will rise if I go through the process (experience) of 'making an effort to lift it'". In the last two cases the word "can" characterized what we should call the expression of a conjecture. (Of course I don't mean that we call the sentence a conjecture because it contains the word "can"; but in calling a sentence a conjecture we referred to the role which the sentence played in the language game; and we translate a word our tribe uses by "can" if "can" is the word we should use under the circumstances described.) Now it is clear that the use of "can" in 59), 60), 61) is closely related to the use of "can" in 46) to 49); differing, however in this, that in 46) to 49) the sentences saying that something *could* happen were not expressions of conjecture. Now one might object to this by saying: Surely we are only willing to use the word "can" in such cases as 46) to 49) because it is reasonable to conjecture in these cases what a man will do in the future from the tests he has passed or from the state he is in.

Now it is true that I have deliberately made up the cases 46) to 49) so as to make a conjecture of this kind seem reasonable.

But I have also deliberately made them up so as *not* to contain a conjecture. We can, if we like, make the hypothesis that the tribe would never use such a form of expression as that used in 49), etc. if experience had not shown them that . . . etc. But this is an assumption which, though possibly correct, is in no way presupposed in the games 46) to 49) as I have actually described them.

62). Let the game be this: A writes down a row of numbers. B watches him and tries to find a system in the sequence of these numbers. When he has done so he says: "Now I can go on". This example is particularly instructive because 'being able to go on' here seems to be something setting in suddenly in the form of a clearly outlined event.—Suppose then that A had written down the row 1, 5, 11, 19, 29. At that point B shouts "Now I can go on". What was it that happened when suddenly he saw how to go on? A great many different things might have happened. Let us assume then that in the present case, while A wrote one number after the other, B busied himself with trying out several algebraic formulae to see whether they fitted. When A had written "19" B had been led to try the formula $a_n = n^2 + n - 1$. A's writing 29 confirms his guess.

63). Or, no formula came into B's mind. After looking at the growing row of numbers A was writing, possibly with a feeling of tension and with hazy ideas floating in his mind, he said to himself the words "He's squaring and always adding one more"; then he made up the next number of the sequence and found it to agree with the numbers A then wrote down.—

64). Or, the row A wrote down was 2, 4, 6, 8. B looks at it, and says "Of course I can go on", and continues the series of even numbers. Or he says nothing, and just goes on. Perhaps when looking at the row 2, 4, 6, 8 which A had written down, he had some sensation, or sensations, often accompanying such words as "That's easy!" A sensation of this kind is, for instance, the experience of a slight, quick intake of breath, what one might call a slight start.

Now, should we say that the proposition "B can continue the series", means that one of the occurrences just described takes place? Isn't it clear that the statement "B can continue . . ." is not the same as the statement that the formula $a_n = n^2 + n - 1$ comes into B's mind? This occurrence might have been all that actually took place. (It is clear, by the way, that it can make no difference to us here whether B has the experience of this formula appearing before his mind's eye, or the experience of writing or speaking the formula, or of picking it out with his eyes from amongst several formulae written down beforehand.) If a parrot had uttered the formula, we should not have said that he could continue the series.—Therefore, we are inclined to say "to be able to . . ." must mean more than just uttering the formula—and in fact more than any one of the occurrences we have described. And this, we go on, shows that saying the formula was only a symptom of B's being able to go on, and that it was not the ability of going on itself. Now what is misleading in this is that we seem to intimate that there is one peculiar activity, process, or state called "being able to go on" which somehow is hidden from our eyes but manifests itself in those occurrents which we call symptoms (as an inflammation of the mucous membranes of the nose produces the symptom of sneezing). This is the way talking of symptoms, in this case, misleads us. When we say "Surely there must be something else behind the mere uttering of the formula, as this alone we should not call 'being able to . . .'", the word "behind" here is certainly used metaphorically, and 'behind' the utterance of the formula may be the circumstances under which it is uttered. It is true, "B can continue . . ." is not the same as to say "B says the formula . . .", but it doesn't follow from this that the expression "B can continue . . ." refers to an activity other than that of saying the formula, in the way in which "B says the formula" refers to the well-known activity. The error we are in is analogous to this: Someone is told the word "chair" does not mean this particular chair I am pointing to, upon which he looks round the room for the object which the word "chair" does denote. (The case would be even more a striking illustration if he tried to look inside the chair in order to find the real meaning of the word "chair".) It is clear that when with reference to the act of writing or speaking the formula etc., we use the sentence "He can continue the series", this must be because of some connection between writing down a formula and actually continuing the series. And the connection in experience of these two processes or activities is clear enough. But this connection tempts us to suggest that the sentence "B can continue . . ." means something like "B does something which, experience has shown us, generally leads to his continuing the series". But does B, when he says "Now I can go on" really mean "Now I am doing something which, as experience has shown us, etc., etc."? Do you mean that he had this phrase in his mind or that he would have been prepared to give it as an explanation of what he had said? To say the phrase "B can continue . . ." is correctly used when prompted by such occurrences as described in 62), 63), 64) but that these occurrences justify its use only under certain circumstances (e.g. when experience has shown certain connections) is not to say that the sentence "B can continue . . ." is short for the sentence which describes all these circumstances, i.e. the whole situation which is the background of our game.

On the other hand we should *under certain circumstances* be ready to substitute "B knows the formula", "B has said the formula" for "B can continue the series". As when we ask a doctor "Can the patient walk?", we shall sometimes be ready to substitute for this "Is his leg healed?"—"Can he speak?" under circumstances means "Is his throat all right?", under others (e.g., if he is a small child) it means "Has he learned to speak?"—To the question "Can the patient walk?", the doctor's answer may be "His leg is all right".—We use the phrase "He can walk, as far as the state of his leg is concerned", especially when we wish to oppose this condition for his walking to some other condition, say the state of his spine. Here we must beware of thinking that there is in the nature of the case something which we might call the complete set of conditions, e.g., for his walking; so that the patient, as it were, can't help walking, *must* walk, if all these conditions are fulfilled.

We can say: The expression "B can continue the series" is used under different circumstances to make different distinctions. Thus it may distinguish *a*) between the case when a man knows the formula and the case when he doesn't; or *b*) between the case when a man knows the formula and hasn't forgotten how to write the numerals of the decimal system, and the case when he knows the formula and has forgotten how to write the numerals; or *c*) (as perhaps in 64)) between the case when a man is feeling his normal self and the case when he is still in a condition of shell-shock; or *d*) between the case of a man who has done this kind of exercise before and the case of a man who is new at it. These are only a few of a large family of cases.

The question whether "He can continue . . ." means the same as "He knows the formula" can be answered in several different ways: We can say "They don't mean the same, i.e., they are not in general used as synonyms as, e.g., the phrases 'I am well' and 'I am in good health'"; or we may say *Under certain circumstances* 'He can continue . . .' means he knows the formula". Imagine the case of a language (somewhat analogous to 49)) in which two forms of expression, two different sentences, are used to say that a person's legs are in working order. The one form of expression is exclusively used under circumstances when preparations are going on for an expedition, a walking tour, or the like; the other is used in cases when there is no question of such preparations. We shall here be doubtful whether to say the two sentences have the same meaning or different meanings. In any case the true state of affairs can only be seen when we look into the detail of the usage of our expressions.—And it is clear that if in our present case we should decide to say that the two expressions have different meanings, we shall certainly not be able to say that the difference is that the fact which makes the second sentence true is a different one from the fact which makes the first sentence true.

We are justified in saying that the sentence "He can continue . . ." has a different meaning from this: "He knows the formula". But we mustn't imagine that we can find a particular state of affairs 'which the first sentence refers to', as it were on a plane above that on which the special occurrences (like knowing the formula, imagining certain further terms, etc.) take place.

Let us ask the following question: Suppose that, on one ground or another, B has said "I can continue the series", but on being asked to continue it he had shown himself unable to do so——should we say that this proved that his statement, that he could continue, was wrong, or should we say that he was able to continue when he said he was? Would B himself say "I see I was wrong", or "What I said was true, I could do it then but I can't now"?—There are cases in which he would correctly say the one and cases in which he would correctly say the other. Suppose *a*) when he said he could continue he saw the formula before his mind, but when he was asked to continue he found he had forgotten it;—or, *b*) when he said he could continue he had said to himself the next five terms of the series, but now finds that they don't come into his mind;—or, *c*) before, he had continued the series calculating five more places, now he still remembers these five numbers but has forgotten how he had calculated them;—or, *d*) he says "Then I felt I could continue, now I can't";—or, *e*) "When I said I could lift the weight my arm didn't hurt, now it does"; etc.

On the other hand we say "I thought I could lift this weight, but I see I can't", "I thought I could say this piece by heart, but I see I was mistaken".

These illustrations of the use of the word "can" should be supplemented by illustrations showing the variety of uses we make of the terms "forgetting" and "trying", for these uses are closely connected with those of the word "can". Consider these cases: *a*) Before, B had said to himself the formula, now, "he finds a complete blank there". *b*) Before, he had said to himself the formula, now, for a moment he isn't sure whether it was 2^n or 3^n, *c*) He has forgotten a name and it is 'on the tip of his tongue'. Or, *d*) he is not certain whether he has ever known the name or has forgotten it.

Now look at the way in which we use the word "trying": *a*) A man is trying to open a door by pulling as hard as he can. *b*) He is trying to open the door of a safe by trying to find the combination. *c*) He is trying to find the combination by trying to remember it, or *d*) by turning the knobs and listening with a stethoscope. Consider the various processes we call "trying to remember". Compare *e*) trying to move your finger against a resistance (e.g., when someone is holding it), and *f*) when you have intertwined the fingers of both hands in a particular way and feel 'you don't know what to do in order to make a particular finger move'.

(Consider also the class of cases in which we say "I can do so and so but I won't": "I could if I tried"—e.g., lift 100 pounds; "I could if I wished"—e.g., say the alphabet.)

One might perhaps suggest that the only case in which it is correct to say, without restriction, that I can do a certain thing, is that in which while saying that I can do it, I actually do it, and that otherwise I ought to say "I can do it as far as . . . is concerned". One may be inclined to think that only in the above case has a person given a real proof of being able to do a thing.

65). But if we look at a language game in which the phrase "I can . . ." is used in this way (i.e., a game in which doing a thing is taken as the only justification for saying that one is able to do it), we see that there is not the *metaphysical* difference between this game and one in which other justifications are accepted for saying "I can do so and so". A game of the kind 65), by the way, shows us the real use of the phrase "If something happens it certainly can happen"; an almost useless phrase in our language. It sounds as though it had some very clear and deep meaning, but like most of the general philosophical propositions it is meaningless except in very special cases.

66). Make this clear to yourself by imagining a language (similar to 49)) which has two expressions for such sentences as "I am lifting a fifty pound weight"; one expression is used whenever the action is performed as a test (say, before an athletic competition), the other expression is used when the action is not performed as a test.

We see that a vast net of family likenesses connects the cases in which the expressions of possibility, "can", "to be able to", etc. are used. Certain characteristic features, we may say, appear in these cases in different combinations: there is, e.g., the element of conjecture (that something will behave in a certain way in the future); the description of the state of something (as a condition for its behaving in a certain way in the future); the account of certain tests someone or something has passed.—

There are, on the other hand, various reasons which incline us to look at the fact of something being possible, someone being able to do something, etc., as the fact that he or it is in a particular state. Roughly speaking, this comes to saying that "A is in the state of being able to do something" is the form of representation we are most strongly tempted to adopt; or, as one could also put it, we are strongly inclined to use the metaphor of something being in a peculiar state for saying that something can behave in a particular way. And this way of representation, or this metaphor, is embodied in the expressions "He is capable of . . .", "He is able to multiply large numbers in his head", "He can play chess": in these sentences the verb is used in the *present tense*, suggesting that the phrases are descriptions of states which exist at the moment when we speak.

The same tendency shows itself in our calling the ability to solve a mathematical problem, the ability to enjoy a piece of music, etc., certain states of the mind; we don't mean by this expression 'conscious mental phenomena'. Rather, a state of the mind in this sense is the state of a hypothetical mechanism, a mind model meant to explain the conscious mental phenomena. (Such things as unconscious or subconscious mental states are features of the mind *model*.) In this way also we can hardly help conceiving of memory as of a kind of storehouse. Note also how sure people are that to the ability to add or to multiply or to say a poem by heart, etc., there *must* correspond a peculiar state of the person's brain, although on the other hand they know next to nothing about such psycho-physiological correspondences. We regard these phenomena as manifestations of this mechanism, and their possibility is the particular construction of the mechanism itself.

Now looking back to our discussion of 43), we see that it was no real explanation of B's being guided by the signs when we said that B was guided if he *could* also have carried out orders consisting in other combinations of dots and dashes than those of 43). In fact, when we considered the question whether B in 43) was guided by the signs, we were all the time inclined to say some such thing as that we could only decide this question with certainty if we could look into the actual mechanism connecting seeing the signs with acting according to them. For we have a definite picture of what in a mechanism we should call certain parts being guided by others. In fact, the mechanism which immediately suggests itself when we wish to show what in such a case as 43) we should call "being guided by the signs" is a mechanism of the type of a pianola. Here, in the working of the pianola we have a clear case of certain actions, those of the hammers of the piano, being guided by the pattern of the holes in the pianola roll. We could use the expression "The pianola is *reading off* the record made by the perforations in the roll", and we might call patterns of such perforations *complex signs* or *sentences*, opposing their function in a pianola to the function which similar devices have in mechanisms of a different type, e.g., the combination of notches and teeth which form a key bit. The bolt of a lock is caused to slide by this particular combination, but we should not say that the movement of the bolt was guided by the way in which we combined teeth and notches, i.e., we should not say that the bolt moved *according* to the pattern of the key bit. You see here the connection between the idea of being guided and the idea of being able to read new combinations of signs; for we should say that the pianola *can* read *any* pattern of perforations, of a particular kind, it is not built for one particular tune or set of tunes (like a musical box),—whereas the bolt of the lock reacts to that pattern of the key bit only which is predetermined in the construction of the lock. We could say that the notches and teeth forming a key bit are not comparable to the words making up a sentence but to the letters making up a word, and that the pattern of the key bit in this sense did not correspond to a complex sign, to a sentence, but to a word.

It is clear that although we might use the ideas of such mechanisms as similes for describing the way in which B acts in the games 42) and 43), no such mechanisms are actually involved in these games. We shall have to say that the use which we made of the expression "to be guided" in our examples of the pianola and of the lock is only one use within a family of usages, though these examples may serve as metaphors, ways of representation, for other usages.

Let us study the use of the expression "to be guided" by studying the use of the word "reading". By "reading" I here mean the activity of translating script into sounds, also of writing according to dictation or of copying in writing a page of print, and suchlike; reading in this sense does not involve any such thing as understanding what you read. The use of the word "reading" is, of course, extremely familiar to us in the circumstances of our ordinary life (it would be extremely difficult to describe these circumstances even roughly). A person, say an Englishman, has as a child gone through one of the normal ways of training in school or at home, he has learned to read his language, later on he reads books, newspapers, letters, etc. What happens when he reads the newspaper?—His eyes glide along the printed words, he pronounces them aloud or to himself, but he pronounces certain words just taking their pattern in as a whole, other words he pronounces after having seen their first few letters only, others again he reads out letter by letter. We should also say that he had read a sentence if while letting his eyes glide along it he had said nothing aloud or to himself, but on being asked afterwards what he had read he was able to reproduce the sentence verbatim or in slightly different words. He may also act as what we might call a mere reading machine, I mean, paying no attention to what he spoke, perhaps concentrating his attention on something totally different. We should in this case say that he read if he acted faultlessly like a reliable machine.—Compare with this case the case of a beginner. He reads the words by spelling them out painfully. Some of the words, however, he just guesses from their contexts, or possibly he knows the piece by heart. The teacher then says that he is pretending to read the words, or just that he is not really reading them. If, looking at this example, we asked ourselves what reading was, we should be inclined to say that it was a particular conscious mental act. This is the case in which we say "Only he knows whether he is reading; nobody else can really know it". Yet we must admit that as far as the reading of a particular word goes, exactly the same thing might have happened in the beginner's mind when he 'pretended' to read as what happened in the mind of the fluent reader when he read the word. We are using the word "reading" in a different way when we talk about the accomplished reader on the one hand and the beginner when a particular conscious process of spelling out the word takes place in a person's mind from the case in which this does not happen.—Then, at least the person who is reading could say that such and such a word was the first which he actually read.—Also, in the different case of a reading machine which is a mechanism connecting signs with the reactions to these signs, e.g., a pianola, we could say "Only after such and such a thing had been done to the machine, e.g., certain parts had been connected by wires, the machine actually read; the first letter which it read was a *d*".—

In the case 67), by calling certain creatures "reading machines" we meant only that they react in a particular way to seeing printed signs. No connection between seeing and reacting, no internal mechanism enters into this case. It would be absurd if the trainer had answered to the question whether he read the word "table" or not, "Perhaps he read it", for there is no doubt in this case about what he actually did. The change which took place was one which we might call a change in the general behaviour of the pupil, and we have in this case not given a meaning to the expression "the first word in the new era". (Compare with this the following case:

. . . .

In our figure a row of dots with large intervals succeeds a row of dots with small intervals. Which is the last dot in the first sequence and which the first dot in the second? Imagine our dots were holes in the revolving disc of a siren. Then we should hear a tone of low pitch following a tone of high pitch (or vice versa). Ask yourself: At which moment does the tone of low pitch begin and the other end?)

There is a great temptation on the other hand to regard the conscious mental act as the only real criterion distinguishing reading from not reading. For we are inclined to say "Surely a man always knows whether he is reading or pretending to read", or "Surely a man always knows when he is really reading". If A tries to make B believe that he is able to read Cyrillic script, cheating him by learning a Russian sentence by heart and then saying it while looking at the printed sentence, we may certainly say that A knows that he is pretending and that his not reading in this case is characterized by a particular personal experience, namely, that of saying the sentence by heart. Also, if A makes a slip in saying it by heart, this experience will be different from that which a person has who makes a slip in *reading*.

68). But supposing now that a man who could read fluently and who was made to read sentences which he had never read before read these sentences, but all the time with the peculiar feeling of knowing the sequence of words by heart. Should we in this case say that he was not reading, i.e., should we regard his personal experience as the criterion distinguishing between reading and not reading?

69). Or imagine this case: A man under the influence of a certain drug is shown a group of five signs, not letters of an existing alphabet; and looking at them with all the outward signs and personal experiences of spelling out a word, pronounces the word "ABOVE". (This sort of thing happens in dreams. After waking up we then say, "It seemed to me that I was reading these signs though they weren't really signs at all".) In such a case some people might be inclined to say that he is reading, others that he isn't. We could imagine that after he had spelt out the word "above" we showed him other combinations of the five signs and that he read them consistently with his reading of the first permutation of signs shown to him. By a series of similar tests we might find that he used what we might call an imaginary alphabet. If this was so, we should be more ready to say "He is reading" than "He imagines that he reads, but he doesn't really".

Note also that there is a continuous series of intermediary

cases between the case when a person knows by heart what is in print before him, and the case in which he spells out the letters of every word without any such help as guessing from the context, knowing by heart, and such like.

Do this: Say by heart the series of cardinals from one to twelve.—Now look at the dial of your watch and *read* this sequence of numbers. Ask yourself what in this case you called reading, that is, what did you do to make it reading?

Let us try this explanation: A person reads if he *derives* the copy which he is producing from the model which he is copying. (I will use the word "model" to mean that which he is reading off, e.g., the printed sentences which he is reading or copying in writing, or such signs as "− − · · −" in 42) and 43) which he is "reading" by his movements, or the scores which a pianist plays off, etc. The word "copy" I use for the sentence spoken or written from the printed one, for the movements made according to such signs as "− − · · −", for the movements of the pianist's fingers or the tune which he plays from the scores, etc.) Thus if we had taught a person the Cyrillic alphabet and had taught him how each letter was pronounced, if then we gave him a piece printed in the Cyrillic script and he spelt it out according to the pronunciation of each letter as we had taught it, we should undoubtedly say that he was deriving the sound of every word from the written and spoken alphabet taught him. And this also would be a clear case of reading. (We might use the expression, "We have taught him the *rule* of the alphabet".)

But let us see; what made us say that he *derived* the spoken words from the printed by means of the rule of the alphabet? Isn't all we know that we told him that this letter was pronounced this way, that letter that way, etc., and that he afterwards read out words in the Cyrillic script? What suggests itself to us as an answer is that he must have shown somehow that he did actually make the transition from the printed to the spoken words by means of the rule of the alphabet which we had given him. And what we mean by his showing this will certainly get clearer if we alter our example and:

70)　assume that he reads off a text by transcribing it, say, from block letters into cursive script. For in this case we can assume the rule of the alphabet to have been given in the form of a table which shows the block alphabet and the cursive alphabet in parallel columns. Then the *deriving* the copy from the text we should imagine this way: The person who copies looks up the table for each letter at frequent intervals, or he says to himself such things as, "Now what's a small *a* like?", or he tries to visualize the table, refraining from actually looking at it.—

71).　But what if, doing all this, he then transcribed an "A" into a "b", a "B" into a "c", and so on? Should we not call this "reading", "deriving", too? We might in this case describe his procedure by saying that he used the table as we should have used it had we not looked straight from left to right like this:

but like this:

though he actually when looking up the table passed with his eyes or finger horizontally from left to right.—But let us suppose now

72)　that going through the normal process of looking up, he transcribed an "A" into an "n", a "B" into an "x", in short, acted, as we might say, according to a scheme of arrows which showed

seems to trail off into nothing? In case 70) the meaning of "deriving" stood out quite clearly, but we told ourselves that this was only one special case of deriving. It seemed to us that the essence of the process of deriving was here presented in a particular dress and that by stripping it of this we should get at the essence. Now in 71), 72), 73) we tried to strip our case of what had seemed but its peculiar costume only to find that what had seemed mere costumes were the essential features of the case. (We acted as though we had tried to find the real artichoke by stripping it of its leaves.) The use of the word "deriving" is indeed exhibited in 70), i.e., this example showed us one of the family of cases in which this word is used. And the explanation of the use of this word, as that of the use of the word "reading" or "being guided by symbols", essentially consists in describing a selection of examples exhibiting characteristic features, some examples showing these features in exaggeration, others showing transitions, certain series of examples showing the trailing off of such features. Imagine that someone wished to give you an idea of the facial characteristics of a certain family, the So and so's, he would do it by showing you a set of family portraits and by drawing your attention to certain characteristic features, and his main task would consist in the proper *arrangement* of these pictures, which, e.g., would enable you to see how certain influences gradually changed the features, in what characteristic ways the members of the family aged, what features appeared more strongly as they did so.

It was not the function of our examples to show us the essence of 'deriving', 'reading', and so forth through a veil of inessential features; the examples were not descriptions of an outside letting us guess at an inside which for some reason or other could not be shown in its nakedness. We are tempted to think that our examples are *indirect* means for producing a certain image or idea in a person's mind,—that they *hint* at something which they cannot show. This would be so in some such case as this: Suppose I wish to produce in someone a mental image of the inside of a particular eighteenth-century room which he is prevented from entering. I therefore adopt this method: I show him the house from the outside, pointing out the windows of the room in question, I further lead him into other rooms of the same period.—

Our method is *purely descriptive*; the descriptions we give are not hints of explanations.

II

1. Do we have a feeling of familiarity whenever we look at familiar objects? Or do we have it usually?

When do we actually have it?

It helps us to ask: What do we contrast the feeling of familiarity with?

One thing we contrast it with is surprise.

One could say: "Unfamiliarity is much more of an experience than familiarity".

We say: A shows B a series of objects. B is to tell A whether the object is familiar to him or not. *a*) The question may be "Does B know what the objects are?" or *b*) "Does he recognize the particular object?"

1). Take the case that B is shown a series of apparatus—a balance, a thermometer, a spectroscope, etc.

2). B is shown a pencil, a pen, an inkpot, and a pebble. Or:

3). Besides familiar objects he is shown an object of which he says "That looks as though it served some purpose, but I don't know what purpose".

   What happens when B recognizes something as a pencil?

   Suppose A had shown him an object looking like a stick. B handles this object, suddenly it comes apart, one of the parts being a cap, the other a pencil. B says "Oh, this is a pencil". He has recognized the object as a pencil.

4). We could say "B always knew what a pencil looked like; he could, e.g., have drawn one on being asked to. He didn't know that the object he was given contained a pencil which he could have drawn any time". Compare with this case 5):

5). B is shown a word written on a piece of paper held upside down. He does not recognize the word. The paper is gradually turned round until B says "Now I see what it is. It is 'pencil'".

We might say "He always knew what the word 'pencil' looked like. He did not know that the word he was shown would, when turned round, look like 'pencil'".

In both cases 4) and 5) you might say something was hidden. But note the different application of "hidden".

6). Compare with this: You read a letter and can't read one of its words. You guess what it must be from the context, and now can read it. You recognize this scratch as an *e*, the second as an *a*, the third as a *t*. This is different from the case where the word "eat" was covered by a blotch of ink, and you only guessed that the word "eat" must have been in this place.

7). Compare: You see a word and can't read it. Someone alters it slightly by adding a dash, lengthening a stroke, or such like. Now you can read it. Compare this alteration with the turning in 5), and note that there is a sense in which while the word was turned round you saw that it was *not* altered. I.e., there is a case in which you say "I looked at the word while it was turned, and I know that it is the same now as it was when I didn't recognize it".

8). Suppose the game between A and B just consisted in this, that B should say whether he knows the object or not but does not say what it is. Suppose he was shown an ordinary pencil, after having been shown a hygrometer which he had never seen before. On being shown the hygrometer he said that he was not familiar with it, on being shown the pencil, that he knew it. What happened when he recognized it?

Must he have told himself, though he didn't tell A, that what he saw was a pencil? Why should we assume this?

Then, when he recognized the pencil, what did he recognize it as?

9). Suppose even that he had said to himself "Oh, this is a pencil", could you compare this case with 4) or 5)? In these cases one might have said "He recognized this as that" (pointing, e.g., for "this" to the covered up pencil and for "that" to an ordinary pencil, and similarly in 5)).

In 8) the pencil underwent no change and the words "Oh, this is a pencil" did not refer to a paradigm, the similarity of which with the pencil shown B had recognized.

Asked "What is a pencil?", B would not have pointed to another object as the paradigm or sample, but could straight away have pointed to the pencil shown to him.

"But when he said 'Oh, this is a pencil', how did he know that it was if he didn't recognize it as something?"—This really comes to saying "How did he recognize 'pencil' as the name of this sort of thing?" Well, how did he recognize it? He just reacted in this particular way by saying this word.

10). Suppose someone shows you colours and asks you to name them. Pointing to a certain object you say "This is red". What would you answer if you were asked "How do you know that this is red?"?

Of course there is the case in which a general explanation was given to B, say, "We shall call 'pencil' anything that one can easily write with on a wax tablet." Then A shows B amongst other objects a small pointed object, and B says "Oh, this is a pencil", after having thought "One could write with this quite easily". In this case, we may say, a *derivation* takes place. In 8), 9), 10) there is no derivation. In 4) we might say that B derived that the object shown to him was a pencil by means of a paradigm, or else no such derivation might have taken place.

Now should we say that B on seeing the pencil after seeing instruments which he didn't know had a feeling of familiarity? Let us imagine what really might have happened. He saw a pencil, smiled, felt relieved, and the name of the object he saw came into his mind or mouth.

Now isn't the feeling of relief just that which characterizes the experience of passing from unfamiliar to familiar things?

2.   We say we experience tension and relaxation, relief, strain and rest in cases as different as these: A man holds a weight with outstretched arm; his arm, his whole body is in a state of tension. We let him put down the weight, the tension relaxes. A man runs, then rests. He thinks hard about the solution of a problem in Euclid, then finds it, and relaxes. He tries to remember a name, and relaxes on finding it.

What if we asked "What do all these cases have in common that makes us say that they are cases of strain and relaxation?"?

What makes us use the expression "seeking in our memory", when we try to remember a word?

Let us ask the question "What is the similarity between looking for a word in your memory and looking for my friend in the park?" What would be the answer to such a question?

One kind of answer certainly would consist in describing a series of intermediate cases. One might say that the case which looking in your memory for something is most similar to is not that of looking for my friend in the park, but, say, that of looking up the spelling of a word in a dictionary. And one might go on interpolating cases. Another way of *pointing out* the similarity would be to say, e.g., "In both these cases at first we can't write down the word and then we can". This is what we call pointing out a common feature.

Now it is important to note that we needn't be aware of such similarities thus pointed out when we are prompted to use the words "seeking", "looking for", etc., in the case of trying to remember.

One might be inclined to say "Surely a similarity must strike us, or we shouldn't be moved to use the same word".—Compare that statement with this: "A similarity between these cases must strike us in order that we should be inclined to use the same picture to represent both". This says that some act must precede the act of using this picture. But why shouldn't what we call "the similarity striking us" consist partially or wholly in our using the same picture? And why shouldn't it consist partially or wholly in our being prompted to use the same phrase?

We say: "This picture (or this phrase) suggests itself to us irresistibly". Well, isn't this an experience?

We are treating here of cases in which, as one might roughly put it, the grammar of a word seems to suggest the 'necessity' of a certain intermediary step, although in fact the word is used in cases in which there is no such intermediary step. Thus we are inclined to say: "A man *must* understand an order before he obeys it", "He must know where his pain is before he can point to it", "He must know the tune before he can sing it", and suchlike.

Let us ask the question: Suppose I had explained to someone the word "red" (or the meaning of the word "red") by having pointed to various red objects and given the ostensive explanation.—What does it mean to say "Now if he has understood the meaning, he will bring me a red object if I ask him to"? This seems to say: If he has really got hold of what is in common between all the objects I have shown him, he will be in the position to follow my order. But what is it that is in common to these objects?

Could you tell me what is in common between a light red and a dark red? Compare with this the following case: I show you two pictures of two different landscapes. In both pictures, amongst many other objects, there is the picture of a bush, and it is exactly alike in both. I ask you "Point to what these two pictures have in common", and as an answer you point to this bush.

Now consider this explanation: I give someone two boxes containing various things, and say "The object which both boxes have in common is called a toasting fork". The person I give this explanation has to sort out the objects in the two boxes until he finds the one they have in common, and thereby, we may say, he arrives at the ostensive explanation. Or, this explanation: "In these two pictures you see patches of many colours; the one colour which you find in both is called 'mauve'".—In this case it makes a clear sense to say "If he has seen (or found) what is in common between these two pictures, he can now bring me a mauve object".

There is also this case: I say to someone "I shall explain to you the word 'w' by showing you various objects. What's in common to them all is what 'w' means." I first show him two books, and he asks himself "Does 'w' mean 'book'?" I then point to a brick, and he says to himself "Perhaps 'w' means 'parallelepiped'". Finally I point to glowing coal, and he says to himself "Oh, it's 'red' he means, for all these objects had something red about them". It would be interesting to consider another form of this game where the person has at each stage to *draw* or *paint* what he thinks I mean. The interest of this version lies in this, that in some cases it would be quite obvious what he has got to draw, say, when he sees that all the objects I have shown him so far bear a certain trademark (he'd draw the trademark).—What, on the other hand, should be paint if he recognizes that there is something red about each object? A red patch? And of what shape and shade? Here a convention would have to be laid down, say, that painting a red patch with ragged edges does not mean that the objects have that red patch with ragged edges in common, but *something* red.

If, pointing to patches of various shades of red, you asked a man "What have these in common that makes you call them red?", he'd be inclined to answer "Don't you see?" And this of course would not be pointing out a common element.

There are cases where experience teaches us that a person is not able to carry out an order, say, of the form "Bring me x" if he did not see what was in common between the various objects to which I pointed as an explanation of "x". And 'seeing what they have in common' in some cases consisted in pointing to it, in letting one's glance rest on a coloured patch after a process of scrutiny and comparing, in saying to oneself "Oh, it's red he means", and perhaps at the same time glancing at all the red patches on the various objects, and so on.—There are cases, on the other hand, in which no process takes place comparable with this intermediary 'seeing what's in common', and where we still use this phrase, though this time we ought to say "If after showing him these things he brings me another red object, then *I shall say* that he has seen the common feature of the objects I showed him". Carrying out the order is now the criterion for his having understood.

3.   Why do you call "strain" all these different experiences?'— 'Because they have some element in common.'—'What is it that bodily and mental strain have in common?'—'I don't know, but obviously there is some similarity.'

Then why did you say the experiences had something in common? Didn't this expression just compare the present case with those cases in which we primarily say that two experiences have something in common? (Thus we might say that some experiences of joy and of fear have the feeling of heart-beat in common.) But when you said that the two experiences of strain had something in common, these were only different words for saying that they were similar. It was then no explanation to say that the similarity consisted in the occurrence of a common element.

Also, shall we say that you had a feeling of similarity when you compared the two experiences, and that this made you use the same word for both? If you say you have a feeling of similarity, let us ask a few questions about it:

Could you say the feeling was located here or there?

When did you actually have this feeling? For, what we call comparing the two experiences is quite a complicated activity: perhaps you called the two experiences before your mind, and imagining a bodily strain, and imagining a mental strain, was each in itself imagining a process and not a state uniform through time. Then ask yourself at what time during all this you had the feeling of similarity.

'But surely I wouldn't say they are similar if I had no experience of their similarity.'—But must this experience be anything you should call a feeling? Suppose for a moment it were the experience that the word "similar" suggested itself to you. Would you call this a feeling?

'But is there no feeling of similarity?'—I think there are feelings which one might call feelings of similarity. But you don't always have any such feeling if you 'notice similarity'. Consider some of the different experiences which you have if you do so.

a) There is a kind of experience which one might call being hardly able to distinguish. You see, e.g., two lengths, two colours, almost exactly alike. But if I ask myself "Does this experience consist in having a peculiar feeling?", I should have to say that it certainly isn't characterized by any such feeling alone, that a most important part of the experience is that of letting my glance oscillate between the two objects, fixing it intently now on the one, now on the other, perhaps saying words expressive of doubt, shaking my head, etc., etc. There is, one might say, hardly any room left for a feeling of similarity between these manifold experiences.

b) Compare with this the case in which it is impossible to have any difficulty in distinguishing the two objects. Supposing I say "I like to have the two kinds of flowers in this bed of similar colours to avoid a strong contrast". The experience here might be one which one may describe as an easy sliding of the glance from one to the other.

c) I listen to a variation on a theme and say "I don't see yet how this is a variation of the theme, but I see a certain similarity". What happened was that at certain points of the variation, at certain turning points of the key, I had an experience of 'knowing where I was in the theme'. And this experience might again have consisted in imagining certain figures of the theme, or in seeing them written before my mind or in actually pointing to them in the score, etc.

'But when two colours are similar, the experience of similarity should surely consist in noticing the similarity which there *is* between them'.—But is a bluish green similar to a yellowish green or not? In certain cases we should say they are similar and in others that they are most dissimilar. Would it be correct to say that in the two cases we noticed different relations between them? Suppose I observed a process in which a bluish green gradually changed into a pure green, into a yellowish green, into yellow, and into orange. I say "It only takes a short time from bluish green to yellowish green, because these colours are similar".—But mustn't you have had some experience of similarity to be able to say this?—The experience may be this, of seeing the two colours and saying that they are both green. Or it may be this, of seeing a band whose colour changes from one end to the other in the way described, and having some one of the experiences which one may call noticing how close to each other bluish green and yellowish green are, compared to bluish green and orange.

We use the word "similar" in a huge family of cases.

There is something remarkable about saying that we use the word "strain" for both mental and physical strain because there is a similarity between them. Should you say we use the word "blue" both for light blue and dark blue because there is a similarity between them? If you were asked "Why do you call this 'blue' also?", you would say "Because this *is* blue, too".

One might suggest that the explanation is that in this case you call "blue" what is *in common* between the two colours, and that, if you called "strain" what was in common between the two experiences of strain, it would have been wrong to say "I called them both 'strain' because they had a certain similarity", but that you would have had to say "I used the word 'strain' in both cases because there is a strain present in both".

Now what should we answer to the question "What do light blue and dark blue have in common?"? At first sight the answer seems obvious: "They are both shades of blue". But this is really a tautology. So let us ask "What do these colours I am pointing to have in common?" (Suppose one is light blue, the other dark blue.) The answer to this really ought to be "I don't know what game you are playing". And it depends upon this game whether I should say they had anything in common, and what I should say they had in common.

Imagine this game: A shows B different patches of colours and asks him what they have in common. B is to answer by pointing to a particular primary colour. Thus if A points to pink and orange, B is to point to pure red. If A points to two shades of greenish blue, B is to point to pure green and pure blue, etc. If in this game A showed B a light blue and a dark blue and asked what they had in common, there would be no doubt about the answer. If then he pointed to pure red and pure green, the answer would be that these have nothing in common. But I could easily imagine circumstances under which we should say that they had something in common and would not hesitate to say what it was: Imagine a use of language (a culture) in which there was a common name for green and red on the one hand, and yellow and blue on the other. Suppose, e.g., that there were two castes, one the patrician caste, wearing red and green garments, the other, the plebeian, wearing blue and yellow garments. Both yellow and blue would always be referred to as plebeian colours, green and red as patrician colours. Asked what a red patch and a green patch have in common, a man of our tribe would not hesitate to say they were both patrician.

We could also easily imagine a language (and that means again a culture) in which there existed no common expression for light blue and dark blue, in which the former, say, was called "Cambridge", the latter "Oxford". If you ask a man of this tribe what Cambridge and Oxford have in common, he'd be inclined to say "Nothing".

Compare this game with the one above: B is shown certain pictures, combinations of coloured patches. On being asked what these pictures have in common, he is to point to a sample of red, say, if there is a red patch in both, to green if there is a green patch in both, etc. This shows you in what different ways this same answer may be used.

Consider such an explanation as "I mean by 'blue' what these two colours have in common".—Now isn't it possible that someone should understand this explanation? He would, e.g., on being ordered to bring another blue object, carry out this order satisfactorily. But perhaps he will bring a red object and we shall be inclined to say: "He seems to notice some sort of similarity between samples we showed him and that red thing".

Note: Some people when asked to sing a note which we strike for them on the piano, regularly sing the fifth of that note. That makes it easy to imagine that a language might have one name only for a certain note and its fifth. On the other hand we should be embarrassed to answer the question: What do a note and its fifth have in common? For of course it is no answer to say: "They have a certain affinity".

It is one of our tasks here to give a picture of the grammar (the use) of the word "a certain".

To say that we use the word "blue" to mean 'what all these shades of colour have in common' by itself says nothing more than that we use the word "blue" in all these cases.

And the phrase "He sees what all these shades have in common", may refer to all sorts of different phenomena, i.e., all sorts of phenomena are used as criteria for 'his seeing that . . .'. Or all that happens may be that on being asked to bring another shade of blue he carries out our order satisfactorily. Or a patch of pure blue may appear before his mind's eye when we show him the different samples of blue: or he may instinctively turn his head towards some other shade of blue which we haven't shown him for sample, etc., etc.

Now should we say that a mental strain and a bodily strain were 'strains' in the same sense of the word or in different (or 'slightly different') senses of the word?—There are cases of this sort in which we should not be doubtful about the answer.

4.   Consider this case: We have taught someone the use of the words "darker" and "lighter". He could, e.g., carry out such an order as "Paint me a patch of colour darker than the one I am showing you". Suppose now I said to him: "Listen to the five vowels a, e, i, o, u and arrange them in order of their darkness". He may just look puzzled and do nothing, but he may (and some people will) now arrange the vowels in a certain order (mostly i, e, a, o, u). Now one might imagine that arranging the vowels in order of darkness presupposed that when a vowel was sounded a certain colour came before a man's mind, that he then arranged these colours in their order of darkness and told you the corresponding arrangement of the vowels. But this actually need not happen. A person will comply with the order: "Arrange the vowels in their order of darkness", without seeing any colours before his mind's eye.

Now if such a person was asked whether u was 'really' darker than e, he would almost certainly answer some such thing as "It isn't really darker, but it somehow gives me a darker impression".

But what if we asked him "What made you use the word 'darker' in this case at all?"?

Again we might be inclined to say "He must have seen something that was in common both to the relation between two colours and to the relation between two vowels". But if he isn't capable of specifying what this common element was, this leaves us just with the fact that he was prompted to use the words "darker", "lighter" in both these cases.

For, note the word "must" in "He must have seen something . . .". When you said that, you didn't mean that from past experience you conclude that he probably did see something, and that's just why this sentence adds nothing to what we know, and in fact only suggests a different form of words to describe it.

If someone said: "I do see a certain similarity, only I can't describe it", I should say: "This itself characterizes your experience".

Suppose you look at two faces and say "They are similar, but I don't know what it is that's similar about them". And suppose that after a while you said: "Now I know; their eyes have the same shape", I should say "Now your experience of their similarity is different from what it was when you saw similarity and didn't know what it consisted in". Now to the question "What made you use the word 'darker' . . . ?" the answer may be "Nothing made me use the word 'darker',—that is, if you ask me for a *reason* why I use it. I just used it, and what is more, I used it with the same intonation of voice, and perhaps with the same facial expression and gesture, which I should in certain cases be inclined to use when applying the word to colours".—It is easier to see this when we speak of a *deep* sorrow, a *deep* sound, a *deep* well. Some people are able to distinguish between fat and lean days of the week. And their experience when they conceive a day as a fat one consists in applying this word together perhaps with a gesture expressive of fatness and a certain comfort.

But you may be tempted to say: This use of the word and gesture is not their primary experience. First of all they have to conceive the day as fat and then they express this conception by word or gesture.

But why do you use the expression "They have to"? Do you know of an experience in this case which you call "the conception, etc."? For if you don't, isn't it just what one might call a linguistic prejudice that made you say "He had to have a conception before . . . etc."?

Rather, you can learn from this example and from others that there are cases in which we may call a particular experience "noticing, seeing, conceiving that so and so is the case", before expressing it by word or gestures, and that there are other cases in which if we talk of an experience of conceiving at all, we have to apply this word to the experience of using certain words, gestures, etc.

When the man said "u isn't really darker than e . . .", it was essential that he meant to say that the word "darker" was used *in different senses* when one talked of one colour being darker than another and, on the other hand, of one vowel being darker than another.

Consider this example: Suppose we had taught a man to use the words "green", "red", "blue" by pointing to patches of these colours. We had taught him to fetch us objects of a certain colour on being ordered "Bring me something red!", to sort out objects of various colours from a heap, and suchlike. Suppose we now show him a heap of leaves, some of which are a slightly reddish brown, others a slightly greenish yellow, and give him the order "Put the red leaves and the green leaves on separate heaps". It is quite likely that he will upon this separate the greenish yellow leaves from the reddish brown ones. Now should we say that we had here used the words "red" and "green" in the same sense as in the previous cases, or did we use them in different but similar senses? What reasons would one give for adopting the latter view? One could point out that on being asked to paint a red patch, one should certainly not have painted a slightly reddish brown one, and therefore one might say "red" means something different in the two cases. But why shouldn't I say that it had one meaning only but was, of course, used according to the circumstances?

The question is: Do we supplement our statement that the word has two meanings by a statement saying that in one case it had this, in the other that meaning? As the criterion for a word's having two meanings, we may use the fact of there being two explanations given for a word. Thus we say the word "bank" has two meanings; for in one case it means this sort of thing (pointing, say, to a river bank), in the other case that sort of thing (pointing to the Bank of England). Now what I point to here are paradigms for the use of the words. One could not say: "The word 'red' has two meanings because in one case it means this (pointing to a light red), in the other that (pointing to a dark red)", if, that is to say, there had been only one ostensive definition for the word "red" used in our game. One could, on the other hand, imagine a language game in which two words, say "red" and "reddish", were explained by two ostensive definitions, the first showing a dark red object, the second a light red one. Whether two such explanations were given or only one might depend on the natural reactions of the people using the language. We might find that a person to whom we give the ostensive definition, "This is called 'red'" (pointing to one red object) thereupon fetches any red object of whatever shade of red on being ordered: "Bring me something red!" Another person might not do so, but bring objects of a certain range of shades only in the neighbourhood of the shade pointed out to him in the explanation. We might say that this person 'does not see what is in common between all the different shades of red'. But remember please that our only criterion for that is the behaviour we have described.

Consider the following case: B has been taught a use of the words "lighter" and "darker". He has been shown objects of various colours and has been taught that one calls this a darker colour than that, trained to bring an object on being ordered "Bring something darker than this", and to describe the colour of an object by saying that it is darker or lighter than a certain sample, etc., etc. Now he is given the order to put down a series of objects, arranging them in the order of their darkness. He does this by laying out a row of books, writing down a series of names of animals, and by writing down the five vowels in the order u, o, a, e, i. We ask him why he put down that latter series, and he says, "Well, o is lighter than u, and e lighter than o".—We shall be astonished at his attitude, and at the same time admit that there is something in what he says. Perhaps we shall say: "But look, surely e isn't lighter than o in the way this look is lighter than that".—But he may shrug his shoulders and say, "I don't know, but e *is* lighter than o, isn't it?"

We may be inclined to treat this case as some kind of abnormality, and to say, "B must have a different sense, with the help of which he arranges both coloured objects and vowels". And if we tried to make this idea of ours (quite) explicit, it would come to this: The normal person registers lightness and darkness of visual objects on one instrument, and, what one might call the lightness and darkness of sounds (vowels) on another, in the sense in which one might say that we record rays of a certain wave length with the eyes, and rays of another range of wave length with our sense of temperature. B on the other hand, we wish to say, arranges both sounds and colours by the readings of one instrument (sense organ) only (in the sense in which a photographic plate might record rays of a range which we could only cover with two of our senses).

This roughly is the picture standing behind our idea that B must have 'understood' the word "darker" differently from the normal person. On the other hand let us put side by side with this picture the fact that there is in our case no evidence for 'another sense'.—And in fact the use of the word "must" when we say "B must have understood the word differently" already shows us that this sentence (really) expresses our determination to look at the phenomena we have observed after* the picture outlined in this sentence.

'But surely he used "lighter" in a different sense when he said e was lighter than u'.—What does this mean? Are you distinguishing between the sense in which he used the word and his usage of the word? That is, do you wish to say that if someone uses the word as B does, some other difference, say in his mind, must go along with the difference in usage? Or is all you want to say that surely the usage of "lighter" was a different one when he applied it to vowels?

Now is the fact that the usages differ anything over and above what you describe when you point out the particular differences?

What if somebody said, pointing to two patches which I had called red, "Surely you are using the word 'red' in two different ways"?—I should say "This is light red and the other dark red,—but why should I have to talk of two different usages?"—

It certainly is easy to point out differences between that part of the game in which we applied "lighter" and "darker" to coloured objects and that part in which we applied these words to vowels. In the first part there was comparison of two objects by laying them side by side and looking from one to the other, there was painting a darker or lighter shade than a certain sample given; in the second there was no comparison by the eye, no painting, etc. But when these differences are pointed out, we are still free to speak of two parts of the same game (as we have done just now) or of two different games.

'But don't I perceive that the relation between a lighter and a

---

* German "nach", i.e. "according to" or "in the light of".—*Edd*.

darker bit of material is a different one than that between the vowels e and u,—as on the other hand I perceive that the relation between u and e is the same as that between e and i?'—Under certain circumstances we shall in these cases be inclined to talk of different relations, under certain others to talk of the same relation. One might say, "It depends how one compares them".

Let us ask the question "Should we say that the arrows → and ← point in the same direction or in different directions?"—At first sight you might be inclined to say "Of course, in different directions". But look at it this way: If I look into a looking glass and see the reflection of my face, I can take this as a criterion for seeing my own head. If on the other hand I saw it in the back of a head I might say "It can't be my own head I am seeing, but a head looking in the opposite direction". Now this could lead me on to say that an arrow and the reflection of an arrow in a glass have the same direction when they point towards each other, and opposite directions when the head of the one points to the tail end of the other. Imagine the case that a man had been taught the ordinary use of the word "the same" in the cases of "the same colour", "the same shape", "the same length". He had also been taught the use of the word "to point to" in such contexts as "The arrow points to the tree". Now we show him two arrows facing each other, and two arrows one following the other, and ask him in which of these two cases he'd apply the phrase "The arrows point the same way". Isn't it easy to imagine that if certain applications were uppermost in his mind, he would be inclined to say that the arrows → ← point 'the same way'?

When we hear the diatonic scale we are inclined to say that after every seven notes the same note recurs, and, asked why we call it the same note again one might answer "Well, it's a c again". But this isn't the explanation I want, for I should ask "What made one call it a c again?" And the answer to this would seem to be "Well, don't you hear that it's the same note only an octave higher?"—Here, too, we could imagine that a man had been taught our use of the word "the same" when applied to colours, lengths, directions, etc., and that we now played the diatonic scale for him and asked him whether he'd say that he heard the same notes again and again at certain intervals, and we could easily imagine several answers, in particular for instance, this, that he heard the same note alternately after every four or three notes (he calls the tonic, the dominant, and the octave the same note).

If we had made this experiment with two people A and B, and A had applied the expression "the same note" to the octave only, B to the dominant and octave, should we have a right to say that the two hear different things when we play to them the diatonic scale?—If we say they do, let us be clear whether we wish to assert that there must be some other difference between the two cases besides the one we have observed, or whether we wish to make no such statement.

5.   All the questions considered here link up with this problem: Suppose you had taught someone to write down series of numbers according to rules of the form: Always write down a number n greater than the preceding. (This rule is abbreviated to "Add n".) The numerals in this game are to be groups of dashes $|, |\ |, |\ |\ |$, etc. What I call teaching this game, of course, consisted in giving general explanations and doing examples.—These examples are taken from the range, say, between 1 and 85. We now give the pupil the order "Add 1". After some time we observe that after passing 100 he did what we should call adding 2; after passing 300 he does what we should call adding 3. We have him up for this: "Didn't I tell you always to add 1? Look what you have done before you got to 100!"—Suppose the pupil said, pointing to the numbers 102, 104, etc., "Well, didn't I do the same here? I thought this was what you wanted me to do."—You see that it would get us no further here again to say "But don't you see . . . ?", pointing out to him again the rules and examples we

had given to him. We might, in such a case, say that this person naturally understands (interprets) the rule (and examples) we have given as we should understand the rule (and examples) telling us: "Add 1 up to 100, then 2 up to 200, etc."

(This would be similar to the case of a man who did not naturally follow an order given by a pointing gesture by moving in the direction shoulder to hand, but in the opposite direction. And understanding here means the same as reacting.)

'I suppose what you say comes to this, that in order to follow the rule "Add 1" correctly a new insight, intuition, is needed at every step.'—But what does it mean to follow the rule *correctly*? How and when is it to be decided which at a particular point is the correct step to take?—'The correct step at every point is that which is in accordance with the rule as it was *meant*, intended.'—I suppose the idea is this: When you gave the rule "Add 1" and meant it, you meant him to write 101 after 100, 199 after 198, 1041 after 1040, and so on. But how did you do all these acts of meaning (I suppose an infinite number of them) when you gave him the rule? Or is this misrepresenting it? And would you say that there was only one act of meaning, from which, however, all these others, or any one of them, followed in turn? But isn't the point just: 'What does follow from the general rule?' You might say "Surely I knew when I gave him the rule that I meant him to follow up 100 by 101". But here you are misled by the grammar of the word "to know". Was knowing this some mental act by which you at the time made the transition from 100 to 101, i.e., some act like saying to yourself "I want him to write 101 after 100"? In this case ask yourself how many such acts you performed when you gave him the rule. Or do you mean by knowing some kind of disposition?—then only experience can teach us what it was a disposition for.—'But surely if one had asked me which number he should write after 1568, I should have answered "1569".'—I dare say you would, but how can you be sure of it? Your idea really is that somehow in the mysterious act of *meaning* the rule you made the transitions without really making them. You crossed all the bridges before you were there.—This queer idea is connected with a peculiar use of the word "to mean". Suppose our man got to the number 100 and followed it up by 102. We should then say "I *meant* you to write 101". Now the past tense in the word "to mean" suggests that a particular act of meaning had been performed when the rule was given, though as a matter of fact this expression alludes to no such act. The past tense could be explained by putting the sentence into the form "Had you asked me before what I wanted you to do at this stage, I should have said . . .". But it is a hypothesis that you would have said that.

To get this clearer, think of this example: Someone says "Napoleon was crowned in 1804". I ask him "Did you mean the man who won the battle of Austerlitz?" He says "Yes, I meant him".—Does this mean that when he 'meant him', he in some way thought of Napoleon's winning the battle of Austerlitz?—

The expression "The rule meant him to follow up 100 by 101" makes it appear that this rule, as it was meant, *foreshadowed* all the transitions which were to be made according to it. But the assumption of a shadow of a transition does not get us any further, because it does not bridge the gulf between it and the real transition. If the mere words of the rule could not anticipate a future transition, no more could any mental act accompanying these words.

We meet again and again with this curious superstition, as one might be inclined to call it, that the mental act is capable of crossing a bridge before we've got to it. This trouble crops up whenever we try to think about the ideas of thinking, wishing, expecting, believing, knowing, trying to solve a mathematical problem, mathematical induction, and so forth.

It is no act of insight, intuition, which makes us use the rule as we do at the particular point of the series. It would be less confusing to call it an act of decision, though this too is misleading, for nothing like an act of decision must take place, but possibly just an act of writing or speaking. And the mistake which we here and in a thousand similar cases are inclined to make is labelled by the word "to make" as we have used it in the sentence "It is no act of insight which makes us use the rule as we do", because there is an idea that 'something must make us' do what we do. And this again joins on to the confusion between cause and reason. *We need have no reason to follow the rule as we do.* The chain of reasons has an end.

Now compare these sentences: "Surely it is using the rule 'Add 1' in a different way if after 100 you go on to 102, 104, etc." and "Surely it is using the word 'darker' in a different way if after applying it to coloured patches we apply it to the vowels".—I should say: "That depends on what you call a 'different way' ".—

But I should certainly say that *I should* call the application of "lighter" and "darker" to vowels 'another usage of the words'; and I also should carry on the series 'Add 1' in the way 101, 102, etc., but not—or not necessarily—because of some other justifying mental act.

6.　There is a kind of general disease of thinking which always looks for (and finds) what would be called a mental state from which all our acts spring as from a reservoir. Thus one says, "The fashion changes because the taste of people changes". The taste is the mental reservoir. But if a tailor to-day designs a cut of dress different from that which he designed a year ago, can't what is called his change of taste have consisted, partly or wholly, in doing just this?

And here we say "But surely designing a new shape isn't in itself changing one's taste,—and saying a word isn't meaning it,—and saying that I believe isn't believing; there must be feelings, mental acts, going along with these lines and these words".—And the reason we give for saying this is that a man certainly could design a new shape without having changed his taste, say that he believes something without believing it, etc. And this obviously is true. But it doesn't follow that what distinguishes a case of having changed one's taste from a case of not having done so isn't under certain circumstances just designing what one hasn't designed before. Nor does it follow that in cases in which designing a new shape is not the criterion for a change of taste, the criterion must be a change in some particular region of our mind.

That is to say, we don't use the word "taste" as the name of a feeling. To think that we do is to represent the practice of our language in undue simplification. This, of course, is the way in which philosophical puzzles generally arise; and our case is quite analogous to that of thinking that wherever we make a predicative statement we state that the subject has a certain ingredient (as we really do in the case, "Beer is alcoholic").

It is advantageous in treating our problem to consider parallel with the feeling or feelings characteristic for having a certain taste, changing one's taste, meaning what one says, etc., etc., the facial expression (gestures or tone of voice) characteristic for the same states or events. If someone should object, saying that feelings and facial expressions can't be compared, as the former are experiences and the latter aren't, let him consider the muscular, kinaesthetic and tactile experiences bound up with gestures and facial expressions.

7.   Let us then consider the proposition "Believing something cannot merely consist in saying that you believe it, you must say it with a particular facial expression, gesture, and tone of voice". Now it cannot be doubted that we regard certain facial expressions, gestures, etc. as characteristic for the expression of belief. We speak of a 'tone of conviction'. And yet it is clear that this tone of conviction isn't always present whenever we rightly speak of conviction. "Just so", you might say, "this shows that

there is something else, something behind these gestures, etc. which is the real belief as opposed to mere expressions of belief".—"Not at all", I should say, "many different criteria distinguish, under different circumstances, cases of believing what you say from those of not believing what you say". There may be cases where the presence of a sensation other than those bound up with gestures, tone of voice, etc. distinguishes meaning what you say from not meaning it. But sometimes what distinguishes these two is nothing that happens while we speak, but a variety of actions and experiences of different kinds before and after.

To understand this family of cases it will again be helpful to consider an analogous case drawn from facial expressions. There is a family of friendly facial expressions. Suppose we had asked "What feature is it that characterizes a friendly face?" At first one might think that there are certain traits which one might call friendly traits, each of which makes the face look friendly to a certain degree, and which when present in a large number constitute the friendly expression. This idea would seem to be borne out by our common speech, talking of 'friendly eyes', 'friendly mouth', etc. But it is easy to see that the same eyes of which we say they make a face look friendly do not look friendly, or even look unfriendly, with certain other wrinkles of the forehead, lines round the mouth, etc. Why then do we ever say that it is these eyes which look friendly? Isn't it wrong to say that they characterize the face as friendly, for if we say they do so 'under certain circumstances' (these circumstances being the other features of the face) why did we single out the one feature from amongst the others? The answer is that in the wide family of friendly faces there is what one might call a main branch characterized by a certain kind of eyes, another by a certain kind of mouth, etc.; although in the large family of unfriendly faces we meet these same eyes when they don't mitigate the unfriendliness of the expression.—There is further the fact that when we notice the friendly expression of a face, our attention, our gaze, is drawn

without the appropriate feelings or even with the opposite feelings. If we lied in giving the information about the train, we would be likely to have different experiences while we gave it than those which we have in giving truthful information, but the difference here would not consist in the absence of a characteristic feeling, but perhaps just in the presence of a feeling of discomfort.

It is even possible while lying to have quite strong experience of what might be called the characteristic for meaning what one says—and yet under certain circumstances, and perhaps under the ordinary circumstances, one refers to just this experience in saying, "I meant what I said", because the cases in which something might give the lie to these experiences do not come into the question. In many cases therefore we are inclined to say: "Meaning what I say" means having such and such experiences while I say it.

If by "believing" we mean an activity, a process, taking place while we say that we believe, we may say that believing is something similar to or the same as expressing a belief.

8.   It is interesting to consider an objection to this: What if I said "I believe it will rain" (meaning what I say) and someone wanted to explain to a Frenchman who doesn't understand English what it was I believed. Then, you might say, if all that happened when I believed what I did was that I said the sentence, the Frenchman ought to know what I believe if you tell him the exact words I used, or say "Il croit 'It will rain'". Now it is clear that this will not tell him what I believe and consequently, you might say, we failed to convey just that to him which was essential, my real mental act of believing.—But the answer is that even if my words had been accompanied by all sorts of experiences, and if we could have transmitted these experiences to the Frenchman, he would still not have known what I believed. For "knowing what I believe" just doesn't mean: feeling what I do while I say it; just as knowing what I intend with this move in

our game of chess doesn't mean knowing my exact state of mind while I'm making the move. Though, at the same time, in certain cases, knowing this state of mind might furnish you with very exact information about my intention.

We should say that we had told the Frenchman what I believed if we translated my words for him into French. And it *might* be that thereby we told him nothing—even indirectly—about what happened 'in me' when I uttered my belief. Rather, we pointed out to him a sentence which in his language holds a similar position to my sentence in the English language.—Again one might say that, at least in certain cases, we could have told him much more exactly what I believed if he had been at home in the English language, because then, he would have known exactly what happened within me when I spoke.

We use the words "meaning", "believing", "intending" in such a way that they refer to certain acts, states of mind given certain circumstances; as by the expression "checkmating somebody" we refer to the act of taking his king. If on the other hand, someone, say a child, playing about with chessmen, placed a few of them on a chess board and went through the motions of taking a king, we should not say the child had checkmated anyone.—And here, too, one might think that what distinguished this case from real checkmating was what happened in the child's mind.

Suppose I had made a move in chess and someone asked me "Did you intend to mate him?", I answer "I did", and he now asks me "How could you know you did, as all you *knew* was what happened within you when you made the move?", I might answer "Under *these* circumstances this was intending to mate him".

9.   What holds for 'meaning' holds for 'thinking'.—We very often find it impossible to think without speaking to ourselves half aloud,—and nobody asked to describe what happened in this case would ever say that something—the thinking—accompanied the speaking, were he not led into doing so by the pair of verbs

"speaking"/"thinking", and by many of our common phrases in which their uses run parallel. Consider these examples: "Think before you speak!" "He speaks without thinking", "What I said didn't quite express my thought", "He says one thing and thinks just the opposite", "I didn't mean a word of what I said", "The French language uses its words in that order in which we think them".

If anything in such a case can be said to go with the speaking, it would be something like the modulation of voice, the changes in timbre, accentuation, and the like, all of which one might call means of expressiveness. Some of these, like the tone of voice and the accent, nobody for obvious reasons would call the accompaniments of the speech; and such means of expressiveness as the play of facial expression or gestures which can be said to accompany speech, nobody would dream of calling thinking.

10.   Let us revert to our example of the use of "lighter" and "darker" for coloured objects and the vowels. A reason which we should like to give for saying that here we have two different uses and not one is this: 'We don't think that the words "darker", "lighter" actually fit the relation between the vowels, we only feel a resemblance between the relation of the sounds and the darker and lighter colours'. Now if you wish to see what sort of feeling this is, try to imagine that without previous introduction you asked someone "Say the vowels a, e, i, o, u, in the order of their darkness". If I did this, I should certainly say it in a different tone from that in which I should say "Arrange these books in the order of their darkness"; that is, I should say it haltingly, in a tone similar to that of "I wonder if you understand me", perhaps smiling slyly as I say it. And this, if anything, describes my *feeling*.

And this brings me to the following point: When someone asks me "What colour is the book over there?", and I say "Red", and then he asks "What made you call this colour 'red'?", I shall in most cases have to say: "Nothing *makes* me call it red; that is, *no reason*. I just looked at it and said 'It's red'". One is then inclined to say: "Surely this isn't all that happened; for I could look at a colour and say a word and still not name the colour". And then one is inclined to go on to say: "The word 'red' when we pronounce it, naming the colour we look at, *comes in a particular way*". But, at the same time, asked "Can you describe the way you mean?"—one wouldn't feel prepared to give *any* description. Suppose now we asked: "Do you, at any rate, remember that the name of the colour came to you in *that particular way* whenever you named colours on former occasions?"—he would have to admit that he didn't remember a particular way in which this always happened. In fact one could easily make him see that naming a colour could go along with all sorts of different experiences. Compare such cases as these: *a*) I put an iron in the fire to heat it to light red heat. I am asking you to watch the iron and want you to tell me from time to time what stage of *heat* it has reached. You look and say: "It is beginning to get light red". *b*) We stand at a street crossing and I say: "Watch out for the green light. When it comes on, tell me and I'll run across." Ask yourself this question: If in one such case you shout "Green!" and in another "Run!", do these words come in the same way or in different ways? Can one say anything about this in a general way? *c*) I ask you: "What's the colour of the bit of material you have in your hand?" (and I can't see). You think: "Now what does one call this? Is this 'Prussian blue' or 'indigo'?"

Now it is very remarkable that when in a philosophical conversation we say: "The name of a colour comes in a particular way", we don't trouble to think of the many different cases and ways in which such a name comes.—And our chief argument is really that naming the colour is different from just pronouncing a word on some different occasion while looking at a colour. Thus one might say: "Suppose we counted some objects lying on our table, a blue one, a red one, a white one, and a black one,—looking at each in turn we say: 'One, two, three, four'. Isn't it easy to see that something different happens in this case when we pronounce the words than what would happen if we had to tell someone the colours of the objects? And couldn't we, with the same right as before, have said 'Nothing else happens when we say the numerals than just saying them while looking at the objects'?"—Now two answers can be given to this: First, undoubtedly, at least in the great majority of cases, counting the objects will be accompanied by different experiences from naming their colours. And it is easy to describe roughly what the difference will be. In counting we know a certain gesture, as it were, beating the number out with one's finger or by nodding one's head. There is, on the other hand, an experience which one might call "concentrating one's attention on the colour", getting the full impression of it. And these are the sort of things one recalls when one says "It is easy to see that something different happens when we count the objects and when we name their colours". But it is in no way necessary that certain peculiar experiences more or less characteristic for counting take place while we are counting, nor that the peculiar phenomenon of gazing at the colour takes place when we look at the object and name its colour. It is true that the processes of counting four objects and of naming their colours will, in most cases at any rate, be different taken as a whole, and *this* is what strikes us; but that doesn't mean at all that we know that something different happens every time in these two cases when we pronounce a numeral on the one hand and a name of a colour on the other.

When we philosophize about this sort of thing we almost invariably do something of this sort: We repeat to ourselves a certain experience, say, by looking fixedly at a certain object and trying to 'read off' as it were the name of its colour. And it is quite natural that doing so again and again we should be inclined to say, "Something particular happens while we say the word 'blue'". For we are aware of going again and again through the same process. But ask yourself: Is this also the process which we usually go through when on various occasions—not philosophizing—we name the colour of an object?

11.   The problem which we are concerned with we also encounter in thinking about volition, deliberate and involuntary action. Think, say, of these examples: I deliberate whether to lift a certain heavyish weight, decide to do it, I then apply my force to it and lift it. Here, you might say, you have a full-fledged case of willing and intentional action. Compare with this such a case as reaching a man a lighted match after having lit with it one's own cigarette and seeing that he wishes to light his; or again the case of moving your hand while writing a letter, or moving your mouth, larynx, etc. while speaking.—Now when I called the first example a full-fledged case of willing, I deliberately used this misleading expression. For this expression indicates that one is inclined in thinking about volition to regard this sort of example as one exhibiting most clearly the typical characteristic of willing. One takes one's ideas, and one's language, about volition from this kind of example and thinks that they must apply—if not in such an obvious way—to all cases which one can properly call cases of willing.—It is the same case that we have met over and over again: The forms of expression of our ordinary language fit most obviously certain very special applications of the words "willing", "thinking", "meaning", "reading", etc., etc. And thus we might have called the case in which a man 'first thinks and then speaks' the full-fledged case of thinking and the case in which a man spells out the words he is reading the full-fledged case of reading. We speak of an 'act of volition' as different from the action which is willed, and in our first example there are lots of different acts clearly distinguishing this case from one in which all that happens is that the hand and the weight lift: there are the preparations of deliberation and decision, there is the effort of lifting. But where do we find the analogues to these processes in our other examples and in innumerable ones we might have given?

Now on the other hand it has been said that when a man, say, gets out of bed in the morning, all that happens may be this: he deliberates, "Is it time to get up?", he tries to make up his mind, and then suddenly *he finds himself getting up*. Describing it this way emphasizes the absence of an act of volition. Now first: where do we find the prototype of such a thing, i.e., how did we come by the idea of such an act? I think the prototype of the act of volition is the experience of muscular effort.—Now there is something in the above description which tempts us to contradict it; we say: "We don't just 'find', observe, ourselves getting up, as though we were observing someone else! It isn't like, say, watching certain reflex actions. If, e.g., I place myself sideways close to a wall, my wall-side arm hanging down outstretched, the back of the hand touching the wall, and if now keeping the arm rigid I press the back of the hand hard against the wall, doing it all by means of the deltoid muscle, if then I quickly step away from the wall, letting my arm hang down loosely, my arm without any action of mine, of its own accord begins to rise; this is the sort of case in which it would be proper to say, 'I *find* my arm rising'."

Now here again it is clear that there are many striking differences between the case of observing my arm rising in this experiment or watching someone else getting out of bed and the case of finding myself getting up. There is, e.g., in this case a perfect absence of what one might call surprise, also I don't *look* at my own movements as I might look at someone turning about in bed, e.g., saying to myself "Is he going to get up?" There is a difference between the voluntary act of getting out of bed and the involuntary rising of my arm. But there is not one common difference between so-called voluntary acts and involuntary ones, viz., the presence or absence of one element, the 'act of volition'.

The description of getting up in which a man says "I just find myself getting up" suggests that he wishes to say that he *observes* himself getting up. And we may certainly say that an attitude of observing is absent in this case. But the observing attitude again is not one continuous state of mind or otherwise which we are in the whole time while, as we should say, we are observing. Rather, there is a family of groups of activities and experiences which we call observing attitudes. Roughly speaking one might say there are observation-elements of curiosity, observant expectation, surprise, and there are, we should say, facial expressions and gestures of curiosity, of observant expectation, and of surprise; and if you agree that there is more than one facial expression characteristic for each of these cases, and that there can be these cases without any characteristic facial expression, you will admit that to each of these three words a *family* of phenomena corresponds.

12. If I had said "When I told him that the train was leaving at 3.30, believing that it did, nothing happened than that I just uttered the sentence", and if someone contradicted me, saying "Surely this couldn't have been all, as you might 'just say a sentence' without believing it",—my answer should be "I didn't wish to say that there was no difference between speaking, believing what you say, and speaking, not believing what you say; but the pair 'believing'/'not believing' refers to various differences in different cases (differences forming a family), not to one difference, that between the presence and the absence of a certain mental state."

13. Let us consider various characteristics of voluntary and involuntary acts. In the case of lifting the heavy weight, the various experiences of effort are obviously most characteristic for lifting the weight voluntarily. On the other hand, compare with this the case of writing, voluntarily, where in most of the ordinary cases there will be no effort; and even if we feel that the writing tires our hands and strains their muscles, this is not the experience of 'pulling' and 'pushing' which we would call typical voluntary actions. Further compare the lifting of your hand when you lift a weight with it with lifting your hand when, e.g.,

you point to some object above you. This will certainly be regarded as a voluntary act, though the element of effort will most
likely be entirely absent; in fact this raising of the arm to point at
an object is very much like raising the eye to look at it, and here
we can hardly conceive of an effort.—Now let us describe an act
of involuntarily raising your arm. There is the case of our experiment, and this was characterized by the utter absence of muscular strain and also by our observant attitude towards the lifting
of the arm. But we have just seen a case in which muscular strain
was absent, and there are cases in which we should call an action
voluntary although we take an observant attitude towards it. But
in a large class of cases it is the peculiar impossibility of taking an
observant attitude towards a certain action which characterizes it
as a voluntary one. Try, e.g., to observe your hand rising when
you voluntarily raise it. Of course you *see* it rising as you do, say,
in the experiment; but you can't somehow follow it in the same
way with your eye. This might get clearer if you compare two
different cases of following lines on a piece of paper with your
eye; *a*) some irregular line like this:

b) a written sentence. You will find that in *a*) the eye, as it were,
alternately slips and gets stuck, whereas in reading a sentence it
glides along smoothly.

Now consider a case in which we do take up an observant attitude towards a voluntary action, I mean the very instructive
case of trying to draw a square with its diagonals by placing a
mirror on your drawing paper and directing your hand by what
you see by looking at it in the mirror. And here one is inclined to
say that our real *actions*, the ones to which volition *immediately*
applies, are not the movements of our hand but something further back, say, the actions of our muscles. We are inclined to
compare the case with this: Imagine we had a series of levers

before us, through which, by a hidden mechanism, we could direct a pencil drawing on a sheet of paper. We might then be in doubt which levers to pull in order to get the desired movement of the pencil; and we could say that *we deliberately* pulled this particular lever, although we didn't deliberately produce the wrong result that we thereby produced. But this comparison, though it easily suggests itself, is very misleading. For in the case of the levers which we saw before us, there was such a thing as deciding which one we were going to pull before pulling it. But does our volition, as it were, play on a keyboard of muscles, choosing which one it was going to use next?—For some actions which we call deliberate it is characteristic that we, in some sense, 'know what we are going to do' before we do it. In this sense we say that we know what object we are going to point to, and what we might call 'the act of knowing' might consist in looking at the object before we point to it or in describing its position by words or pictures. Now we could describe our drawing the square through the mirror by saying that our acts were deliberate as far as their motor aspect is concerned, but not as far as their visual aspect is concerned. This would, e.g., be demonstrated by our ability to repeat a movement of the hand which had produced a wrong result, on being told to do so. But it would obviously be absurd to say that this motor character of voluntary motion consisted in our knowing beforehand what we were going to do, as though we had had a picture of the kinaesthetic sensation before our mind and decided to bring about this sensation. Remember the experiment where the subject has his fingers intertwined; if here, instead of pointing from a distance to the finger which you order him to move, you touch that finger, he will always move it without the slightest difficulty. And here it is tempting to say "Of course I can move it now, because now I know which finger it is I'm asked to move." This makes it appear as though I had now shown you which muscle to contract in order to bring about the desired result. The word "of course" makes it appear as though by touching your finger I had given you an item of information telling you what to do. (As though normally when you tell a man to move such and such a finger he could follow your order because he knew how to bring the movement about.)

(It is interesting here to think of the case of sucking a liquid through a tube; if asked what part of your body you sucked with, you would be inclined to say your mouth, although the work was done by the muscles by which you draw your breath.)

Let us now ask ourselves what we should call "speaking involuntarily". First note that when normally you speak, voluntarily, you could hardly describe what happened by saying that by an act of volition you move your mouth, tongue, larynx, etc. as a means to producing certain sounds. Whatever happens in your mouth, larynx, etc. and whatever sensations you have in these parts while speaking would almost seem secondary phenomena accompanying the production of sounds, and volition, one wishes to say, operates on the sounds themselves without intermediary mechanism. This shows how loose our idea of this agent 'volition' is.

Now to involuntary speaking. Imagine you had to describe a case,—what would you do? There is of course the case of speaking in one's sleep; this is characterized by our doing it without being aware of it and not remembering having done it. But this obviously you wouldn't call the characteristic of an involuntary action.

A better example of involuntary speaking would, I suppose, be that of involuntary exclamations: "Oh!", "Help!", and such like, and these utterances are akin to shrieking with pain. (This, by the way, could set us thinking about 'words as expressions of feelings'.) One might say: "Surely these are good examples of involuntary speech, because there is in these cases not only no act of volition by which we speak, but in many cases we utter these words *against* our will". I should say: I certainly should call this involuntary speaking; and I agree that an act of volition preparatory to or accompanying these words is absent,—if by "act of volition" you refer to certain acts of intention, premeditation, or effort. But then in many cases of voluntary speech I don't feel an effort, much that I say voluntarily is not premeditated, and I don't know of any acts of intention preceding it.

Crying out with pain against our will could be compared with raising our arm against our will when someone forces it up while we are struggling against him. But it is important to notice that the will—or should we say 'wish'—not to cry out is overcome in a different way from that in which our resistance is overcome by the strength of the opponent. When we cry out against our will, we are as it were taken by surprise; as though someone forced up our hands by unexpectedly sticking a gun into our ribs, commanding "Hands up!"

14.   Consider now the following example, which is of great help in all these considerations: In order to see what happens when one understands a word, we play this game: You have a list of words, partly these words are words of my native language, partly words of foreign languages more or less familiar to me, partly words of languages entirely unknown to me (or, which comes to the same, nonsensical words invented for the occasion). Some of the words of my native tongue, again, are words of ordinary everyday usage: and some of these, like "house", "table", "man", are what we might call primitive words, being among the first words a child learns, and some of these again, words of baby talk like "Mamma", "Papa". Again, there are more or less common technical terms such as "carburettor", "dynamo", "fuse"; etc., etc. All these words are read out to me, and after each one I have to say "Yes" or "No" according to whether I understand the word or not. I then try to remember what happened in my mind when I understood the words I did understand, and when I didn't understand the others. And here again it will be useful to consider the particular tone of voice and facial expression with which I say "Yes" and "No", alongside of the so-called mental

events.—Now it may surprise us to find that although this experiment will show us a multitude of different characteristic experiences, it will not show us any one experience which we should be inclined to call the experience of understanding. There will be such experiences as these: I hear the word "tree" and say "Yes" with the tone of voice and sensation of "Of course". Or I hear "corroboration"——I say to myself, "Let me see", vaguely remember a case of helping, and say "Yes". I hear "gadget", I imagine the man who always used this word, and say "Yes". I hear "Mamma", this strikes me as funny and childish—"Yes". A foreign word I shall very often translate in my mind into English before answering. I hear "spinthariscope", and say to myself, "Must be some sort of scientific instrument", perhaps try to think up its meaning from its derivation and fail, and say "No". In another case I might say to myself, "Sounds like Chinese"— "No". Etc. There will, on the other hand, be a large class of cases in which I am not aware of anything happening except hearing the word and saying the answer. And there will also be cases in which I remember experiences (sensations, thoughts) which, as I should say, had nothing to do with the word at all. Thus amongst the experiences which I can describe there will be a class which I might call typical experiences of understanding and some typical experiences of not understanding. But opposed to these there will be a large class of cases in which I should have to say "I know of no particular experience at all, I just said 'Yes', or 'No'".

Now if someone said "But surely something did happen when you understood the word 'tree', unless you were utterly absent-minded when you said 'Yes'", I might be inclined to reflect and say to myself, "Didn't I have a sort of homely feeling when I took in the word 'tree'?" But then, do I always have this feeling which now I referred to when I hear that word used or use it myself, do I remember having had it, do I even remember a set of, say, five sensations some one of which I had on every occasion when I could be said to have understood the word? Further, isn't that 'homely feeling' I referred to an experience rather characteristic for the particular situation I'm in at present, i.e., that of philosophizing about 'understanding'?

Of course in our experiment we might call saying "Yes" or "No" characteristic experiences of understanding or not understanding, but what if we just hear a word in a sentence, where there isn't even a question of this reaction to it?—We are here in a curious difficulty: on the one hand it seems we have no reason to say that in all cases in which we understand a word one particular experience—or even one of a set—is present. On the other hand we may feel it's plainly wrong to say that in such a case all that happens may be that I hear or say the word. For that seems to be saying that part of the time we act as mere automatons. And the answer is that in a sense we do and in a sense we don't.

If someone talked to me with a kindly play of facial expressions, is it necessary that in any short interval his face should have looked such that seeing it under any other circumstances I should have called its expression distinctly kindly? And if not, does this mean that his 'kindly play of expression' was interrupted by periods of inexpressiveness?—We certainly should not say this under the circumstances which I am assuming, and we don't feel that the look at this moment interrupts the expressiveness, although taken alone we should call it inexpressive.

Just in this way we refer by the phrase "understanding a word" not necessarily to that which happens while we are saying or hearing it, but to the whole environment of the event of saying it. And this also applies to our saying that someone speaks like an automaton or like a parrot. Speaking with understanding certainly differs from speaking like an automaton, but this doesn't mean that the speaking in the first case is all the time accompanied by something which is lacking in the second case. Just as when we say that two people move in different circles this doesn't mean that they mayn't walk the street in identical surroundings.

Thus also, acting voluntarily (or involuntarily) is, in many cases, characterized as such by a multitude of circumstances under which the action takes place rather than by an experience which we should call characteristic of voluntary action. And in this sense it is true to say that what happened when I got out of bed—when I should certainly not call it involuntary—was that I found myself getting up. Or rather, this is a possible case; for of course every day something different happens.

15.   The troubles which we have been turning over since §7 were all closely connected with the use of the word "particular". We have been inclined to say that seeing familiar objects we have a particular feeling, that the word "red" came in a particular way when we recognized the colour as red, that we had a particular experience when we acted voluntarily.

Now the use of the word "particular" is apt to produce a kind of delusion and roughly speaking this delusion is produced by the double usage of this word. On the one hand, we may say, it is used preliminary to a specification, description, comparison; on the other hand, as what one might describe as an emphasis. The first usage I shall call the transitive one, the second the intransitive one. Thus, on the one hand I say "This face gives me a particular impression which I can't describe". The latter sentence may mean something like: "This face gives me a strong impression". These examples would perhaps be more striking if we substituted the word "peculiar" for "particular", for the same comments apply to "peculiar". If I say "This soap has a peculiar smell: it is the kind we used as children", the word "peculiar" may be used merely as an introduction to the comparison which follows it, as though I said "I'll tell you what this soap smells like: . . .". If, on the other hand, I say "This soap has a *peculiar* smell!" or "It has a most peculiar smell", "peculiar" here stands for some such expression as "out of the ordinary", "uncommon", "striking".

We might ask "Did you say it had a peculiar smell, as opposed to no peculiar smell, or that it had this smell, as opposed to some other smell, or did you wish to say both the first and the second?"—Now what was it like when, philosophizing, I said that the word "red" came in a particular way when I described something I saw as red? Was it that I was going to describe the way in which the word "red" came, like saying "It always comes quicker than the word 'two' when I'm counting coloured objects", or "It always comes with a shock", etc.?—Or was it that I wished to say that "red" comes in a striking way?—Not exactly that either. But certainly rather the second than the first. To see this more clearly, consider another example: You are, of course, constantly changing the position of your body throughout the day; arrest yourself in any such attitude (while writing, reading, talking, etc., etc.) and say to yourself in the way in which you say "'Red' comes in a particular way . . .", "I am now in a particular attitude". You will find that you can quite naturally say this. But aren't you always in a particular attitude? And of course you didn't mean that you were just then in a particularly striking attitude. What was it that happened? You concentrated on, as it were stared at, your sensations. And this is exactly what you did when you said that "red" came in a particular way.

"But didn't I mean that 'red' came in a different way from 'two'?"—You may have meant this, but the phrase, "They come in different ways", is itself liable to cause confusion. Suppose I said "Smith and Jones always enter my room in different ways": I might go on and say "Smith enters quickly, Jones slowly", I am specifying the ways. I might on the other hand say "I don't know what the difference is", intimating that I'm *trying* to specify the difference, and perhaps later on I shall say "Now I know what it is; it is . . .".—I could on the other hand tell you that they came in different ways, and you wouldn't know what to make of this statement, and perhaps answer "Of course they come in different ways; they just *are* different".—We could describe our trouble by saying that we feel as though we could give an experience a name without at the same time committing ourselves about its use, and in fact without any intention to use it at all. Thus when I say "red" comes in a particular way . . . , I feel that I might now give this way a name if it hasn't already got one, say "A". But at the same time I am not at all prepared to say that I recognize this to be the way "red" has always come on such occasions, nor even to say that there are, say, four ways, A, B, C, D, in one of which it always comes. You might say that the two ways in which "red" and "two" come can be identified by, say, exchanging the meaning of the two words, using "red" as the second cardinal numeral, "two" as the name of a colour. Thus, on being asked how many eyes I had, I should answer "red", and to the question "What is the colour of blood?", "two". But the question now arises whether you can identify the 'way in which these words come' independently of the ways in which they are used,—I mean the ways just described. Did you wish to say that as a matter of experience, the word when used in *this* way always comes in the way A, but may, the next time, come in the way "two" usually comes? You will see then that you meant nothing of the sort.

What is *particular* about the way "red" comes is that it comes while you're philosophizing about it, as what is particular about the position of your body when you concentrated on it was concentration. We appear to ourselves to be on the verge of describing the way, whereas we aren't really opposing it to any other way. We are emphasizing, not comparing, but we express ourselves as though this emphasis was really a comparison of the object with itself; there seems to be a reflexive comparison. Let me explain myself in this way: suppose I speak of the way in which A enters the room, I may say "I have noticed the way in which A enters the room", and on being asked "What is it?", I may answer "He always sticks his head into the room before coming in". Here I'm referring to a definite feature, and I could say that B had the same way, or that A no longer had it. Consider on the other hand the statement "I've now been observing the way A sits and smokes". I want to draw him like this. In this case I needn't be ready to give any description of a particular feature of his attitude, and my statement may just mean "I've been observing A as he sat and smoked".—'The way' can't in this case be separated from him. Now if I wished to draw him as he sat there, and was contemplating, studying, his attitude, I should while doing so be inclined to say and repeat to myself "He has a particular way of sitting". But the answer to the question "What way?" would be "Well, *this* way", and perhaps one would give it by drawing the characteristic outlines of his attitude. On the other hand, my phrase "He has a particular way . . .", might just have to be translated into "I'm contemplating his attitude". Putting it in this form we have, as it were, straightened out the proposition; whereas in its first form its meaning seems to describe a loop, that is to say, the word "particular" here seems to be used transitively and, more particularly, reflexively, i.e., we are regarding its use as a special case of the transitive use. We are inclined to answer the question "What way do you mean?" by "*This* way", instead of answering: "I didn't refer to any particular feature; I was just contemplating his position". My expression made it appear as though I was pointing out something *about* his way of sitting, or, in our previous case, about the way the word "red" came, whereas what makes me use the word "particular" here is that by my attitude towards the phenomenon I am laying an emphasis on it: I am concentrating on it, or retracing it in my mind, or drawing it, etc.

Now this is a characteristic situation to find ourselves in when thinking about philosophical problems. There are many troubles which arise in this way, that a word has a transitive and an intransitive use, and that we regard the latter as a particular case of the former, explaining the word when it is used intransitively by a reflexive construction.

Thus we say, "By 'kilogram' I mean the weight of one litre of water", "By 'A' I mean 'B', where B is an explanation of A". But there is also the intransitive use: "I said that I was sick of it and meant it". Here again, meaning what you said could be called "retracing it", "laying an emphasis on it". But using the word "meaning" in this sentence makes it appear that it must have sense to ask "*What* did you mean?" and to answer "By what I said I meant what I said," treating the case of "I mean what I say" as a special case of "By saying 'A' I mean 'B'". In fact one uses the expression "I mean what I mean" to say, "I have no explanation for it". The question, "What does this sentence *p* mean?", if it doesn't ask for a translation of *p* into other symbols, has no more sense than "What sentence is formed by this sequence of words?"

Suppose to the question, "What's a kilogram?" I answered, "It is what a litre of water weighs", and someone asked, "Well, what does a litre of water weight?"—

We often use the reflexive form of speech as a means of emphasizing something. And in all such cases our reflexive expressions can be 'straightened out'. Thus we use the expression "If I can't, I can't", "I am as I am", "It is just what it is", also "That's that". This latter phrase means as much as "That's settled", but why should we express "That's settled" by "That's that"? The answer can be given by laying before ourselves a series of interpretations which make a transition between the two expressions. Thus for "That's settled", I will say "The matter is closed". And this expression, as it were, files the matter and shelves it. And filing it is like drawing a line around it, as one sometimes draws a line around the results of a calculation, thereby marking it as final. But this also makes it stand out; it is a way of emphasizing it. And what the expression "That's that" does is to emphasize the 'that'.

Another expression akin to those we have just considered is this: "Here it is; take it or leave it!" And this again is akin to a kind of introductory statement which we sometimes make before remarking on certain alternatives, as when we say: "It either rains or it doesn't rain; if it rains we'll stay in my room, if it doesn't . . .". The first part of this sentence is no piece of information (just as "Take it or leave it" is no order). Instead of, "It either rains or it doesn't rain" we could have said, "Consider the two cases . . .". Our expression underlines these cases, presents them to your attention.

It is closely connected with this that in describing a case like 30)* we are tempted to use the phrase, "There is, *of course*, a number beyond which no one of the tribe has ever counted; let this number be . . .". Straightened out this reads: "Let the number beyond which no one of the tribe has ever counted be . . .". Why we tend to prefer the first expression to the one straightened out is that it more strongly directs our attention to the upper end of the range of numerals used by our tribe in their actual practice.

16. Let us now consider a very instructive case of that use of the word "particular" in which it does not point to a comparison and yet seems most strongly to do so,—the case when we contemplate the expression of a face primitively drawn in this way:

Let this face produce an impression on you. You may then feel inclined to say: "Surely I don't see mere dashes. I see a face with a *particular* expression". But you don't mean that it has an outstanding expression nor is it said as an introduction to a description of the expression, though we might give such a description and say, e.g., "It looks like a complacent business man, stupidly supercilious, who though fat, imagines he's a lady killer". But this would only be meant as an approximate description of the expression. "Words can't exactly describe it", one sometimes says. And yet one feels that what one calls the expression of the face is something that can be detached from the drawing of the face. It is as though we could say: "This face has a particular expression: namely this" (pointing to something). But if I had to point to anything in this place it would have to be the drawing I am looking at. (We are, as it were, under an optical delusion which by some

---

* Language game no. 30 in Part I of the *Brown Book*.

sort of reflection makes us think that there are two objects where there is only one. The delusion is assisted by our using the verb "to have", saying "The face *has* a particular expression". Things look different when, instead of this, we say: "This *is* a peculiar face". What a thing *is*, we mean, is bound up with it; what it has can be separated from it.)

'This face has a particular expression.'—I am inclined to say this when I am trying to let it make its full impression upon me.

What goes on here is an act, as it were, of digesting it, getting hold of it, and the phrase "getting hold of the expression of this face" suggests that we are getting hold of a thing which is *in* the face and different from it. It seems we are looking for something, but we don't do so in the sense of looking for a model of the expression outside the face we see, but in the sense of sounding the thing without attention. It is, when I let the face make an impression on me, as though there existed a double of its expression, as though the double was the prototype of the expression and as though seeing the expression of the face was finding the prototype to which it corresponded—as though in our mind there had been a mould and the picture we see had fallen into that mould, fitting it. But it is rather that we let the picture sink into our mind and make a mould there.

When we say, "This is a *face*, and not mere strokes", we are, of course, distinguishing such a drawing

from such a one

And it is true: If you ask anyone: "What is this?" (pointing to the first drawing) he will certainly say: "It's a face", and he will be able straight away to reply to such questions as, "Is it male or female?", "Smiling or sad?", etc. If on the other hand you ask him: "What is this?" (pointing to the second drawing), he will most likely say, "This is nothing at all", or "These are just dashes". Now think of looking for a man in a picture puzzle; there it often happens that what at first sight appears as 'mere dashes' later appears as a face. We say in such cases: "Now I see it is a face". It must be quite clear to you that this doesn't mean that we recognize it as the face of a friend or that we are under the delusion of seeing a 'real' face: rather, this 'seeing it *as a face*' must be compared with seeing this drawing

either as a cube or as a plane figure consisting of a square and two rhombuses; or with seeing this

'as a square with diagonals', or 'as a swastika', that is, as a limiting case of this;

or again with seeing these four dots . . . . as two pairs of dots side by side with each other, or as two interlocking pairs, or as one pair inside the other, etc.

The case of 'seeing

as a swastika' is of special interest because this expression might mean being, somehow, under the optical delusion that the

square is not quite closed, that there are the gaps which distinguish the swastika from our drawing. On the other hand it is
quite clear that this was not what we meant by "seeing our
drawing as a swastika". We saw it in a way which suggested the
description, "I see it as a swastika". One might suggest that we
ought to have said "I see it as a closed swastika";—but then,
what is the difference between a closed swastika and a square
with diagonals? I think that in this case it is easy to recognize
'what happens when we see our figure as a swastika'. I believe it
is that we retrace the figure with our eyes in a particular way,
viz., by staring at the centre, looking along a radius, and along a
side adjacent to it, starting at the centre again, taking the next
radius and the next side, say in a right-handed sense of rotation,
etc. But this *explanation* of the phenomenon of seeing the figure
as a swastika is of no fundamental interest to us. It is of interest
to us only in so far as it helps one to see that the expression "seeing the figure as a swastika" did not mean seeing *this* or *that*,
seeing one thing as something else, when, essentially, *two* visual
objects entered the process of doing so.—Thus also seeing the
first figure as a cube did not mean 'taking it to be a cube'. (For
we might never have seen a cube and still have this experience of
'seeing it as a cube'.)

And in this way 'seeing dashes as a face' does not involve a
comparison between a group of dashes and a real human face;
and, on the other hand, this form of expression most strongly
suggests that we are alluding to a comparison.

Consider also this example: Look at W once as a capital
double-U, and another time as a capital M upside down. Observe
what doing the one and doing the other consists in.

We distinguish seeing a drawing as a face and seeing it as
something else or as 'mere dashes'. And we also distinguish between superficially glancing at a drawing (seeing it as a face), and
letting the face make its full impression on us. But it would be
queer to say: "I am letting the face make *a particular* impression

on me" (except in such cases in which you can say that you can let the same face make different impressions on you). And in letting the face impress itself on me and contemplating its 'particular impression', no two things of the multiplicity of a face are compared with each other; there is only *one* which is laden with emphasis. Absorbing its expression, I don't find a prototype of this expression in my mind; rather, I, as it were, cut a seal from the impression.

And this also describes what happens when in 15)* we say to ourselves "The word 'red' comes in a particular way . . .". The reply could be: "I see, you're repeating to yourself some experience and again and again gazing at it."

17.   We may shed light on all these considerations if we compare what happens when we remember the face of someone who enters our room, when we recognize him as Mr. So and So,—when we compare what really happens in such cases with the representation we are sometimes inclined to make of the events. For here we are often obsessed by a primitive conception, viz., that we are comparing the man we see with a memory image in our mind and we find the two to agree. I.e., we are representing 'recognizing someone' as a process of identification by means of a picture (as a criminal is identified by his photo). I needn't say that in most cases in which we recognize someone no comparison between him and a mental picture takes place. We are, of course, tempted to give this description by the fact that there are memory images. Very often, for instance, such an image comes before our mind immediately *after* having recognized someone. I see him as he stood when we last saw each other ten years ago.

I will here again describe the *kind* of thing that happens in your mind and otherwise when you recognize a person coming into your room by means of what you might *say* when you recog-

_____

* §15, *Brown Book*, Part II.

nize him. Now this may just be: "Hello!" And thus we may say that one kind of event of recognizing a thing we see consists in saying "Hello!" to it in words, gestures, facial expressions, etc.—And thus also we may think that when we look at our drawing and see it as a face, we compare it with some paradigm, and it agrees with it, or it fits into a mould ready for it in our mind. But no such mould or comparison enters into our experience, there is only this shape, not any other to compare it with, and as it were, say "Of course" to. As when in putting together a jig-saw puzzle, somewhere a small space is left unfilled and I see a piece obviously fitting it and put it in the place saying to myself "Of course". But here we say, "Of course" *because* the piece fits the mould, whereas in our case of seeing the drawing as a face, we have the same attitude for *no* reason.

The same strange illusion which we are under when we seem to seek the something which a face expresses whereas, in reality, we are giving ourselves up to the features before us—that same illusion possesses us even more strongly if repeating a tune to ourselves and letting it make its full impression on us, we say "This tune says *something*", and it is as though I had to find *what* it says. And yet I know that it doesn't say anything such that I might express in words or pictures what it says. And if, recognizing this, I resign myself to saying "It just expresses a musical thought", this would mean no more than saying "It expresses itself".—"But surely when you play it you don't play it *anyhow*, you play it in this particular way, making a crescendo here, a diminuendo there, a caesura in this place, etc."—Precisely, and that's all I can say about it, or may be all that I can say about it. For in certain cases I can justify, explain the particular expression with which I play it by a comparison, as when I say "At this point of the theme, there is, as it were, a colon", or "This is, as it were, the answer to what came before", etc. (This, by the way, shows what a 'justification' and an 'explanation' in aesthetics is like.) It is true I may hear a tune played and say "This is not how it ought to be played, it goes like this"; and I whistle it in a different tempo. Here one is inclined to ask "What is it like to know the tempo in which a piece of music should be played?" And the idea suggests itself that there *must* be a paradigm somewhere in our mind, and that we have adjusted the tempo to conform to that paradigm. But in most cases if someone asked me "How do you think this melody should be played?", I will, as an answer, just whistle it in a particular way, and nothing will have been present to my mind but the tune *actually whistled* (not an image of *that*).

This doesn't mean that suddenly understanding a musical theme may not consist in finding a form of verbal expression which I conceive as the verbal counterpoint of the theme. And in the same way I may say "Now I understand the expression of this face", and what happened when the understanding came was that I found the word which seemed to sum it up.

Consider also this expression: "Tell yourself that it's a *waltz*, and you will play it correctly".

What we call "understanding a sentence" has, in many cases, a much greater similarity to understanding a musical theme than we might be inclined to think. But I don't mean that understanding a musical theme is more like the picture which one tends to make oneself of understanding a sentence; but rather that this picture is wrong, and that understanding a sentence is much more like what really happens when we understand a tune than at first sight appears. For understanding a sentence, we say, points to a reality outside the sentence. Whereas one might say "Understanding a sentence means getting hold of its content; and the content of the sentence is *in* the sentence."

18. We may now return to the ideas of 'recognizing' and 'familiarity', and in fact to that example of recognition and familiarity which started our reflections on the use of these terms and of a multitude of terms connected with them. I mean the example of reading, say, a written sentence in a well-known

language.—I read such a sentence to see what the experience of reading is like, what 'really happens' when one reads, and I get a particular experience which I take to be the experience of reading. And, it seems, this doesn't simply consist in seeing and pronouncing the words, but, besides, in an experience of an intimate character, as I should like to say. (I am as it were on an intimate footing with the word 'I read'.)

In reading the spoken words come in a particular way, I am inclined to say; and the written words themselves which I read don't just look to me like any kind of scribbles. At the same time I am unable to point to, or get a grasp on, that 'particular way'.

The phenomenon of seeing and speaking the words seems enshrouded by a particular atmosphere. But I don't recognize this atmosphere as one which always characterized the situation of reading. Rather, I notice it when I read a line, trying to see what reading is like.

When noticing this atmosphere I am in the situation of a man who is working in his room, reading, writing, speaking, etc., and who suddenly concentrates his attention on some soft uniform noise, such as one can almost always hear, particularly in a town (the dim noise resulting from all the various noises of the street, the sounds of wind, rain, workshops, etc.). We could imagine that this man might think that a particular noise was a common element of all the experiences he had in this room. We should then draw his attention to the fact that most of the time he hadn't noticed any noises going on outside, and secondly, that the noise he could hear wasn't always the same (there was sometimes wind, sometimes not, etc.).

Now we have used a misleading expression when we said that besides the experiences of seeing and speaking in reading there was another experience, etc. This is saying that to certain experiences another experience is added.—Now take the experience of seeing a sad face, say in a drawing,—we can say that to see the drawing as a sad face is not 'just' to see it as some complex of strokes (think of a puzzle picture). But the word 'just' here seems to intimate that in seeing the drawing as a face some experience is added to the experience of seeing it as mere strokes; as though I had to say that seeing the drawing as a face consisted of two experiences, elements.

You should now notice the difference between the various cases in which we say that an experience consists of several elements or that it is a *compound* experience. We might say to the doctor, "I don't have one pain; I have two: toothache and headache". And one might express this by saying, "My experience of pain is not simple, but compound, I have toothache and headache". Compare with this case that in which I say, "I have got both pains in my stomach and a general feeling of sickness". Here I don't separate the constituent experiences by pointing to two localities of pain. Or consider this statement: "When I drink sweet tea, my taste experience is a compound of the taste of sugar and the taste of tea". Or again: "If I hear the C Major chord my experience is composed of hearing C, E, and G". And, on the other hand, "I hear a piano playing and some noise in the street". A most instructive example is this: In a song words are sung to certain notes. In what sense is the experience of hearing the vowel *a* sung to the note C a composite one? Ask yourself in each of these cases: What is it like to single out the constituent experiences in the compound experience?

Now although the expression that seeing a drawing as a face is not merely seeing strokes seems to point to some kind of addition of experiences, we certainly should not say that when we see the drawing as a face we also have the experience of seeing it as mere strokes and some other experience *besides*. And this becomes still clearer when we imagine that someone said that seeing the drawing as a cube consisted in seeing it as a plane figure plus having an experience of depth.

Now when I felt that though while reading a certain constant experience went on and on, I could not in a sense lay hold of that experience, my difficulty arose through wrongly comparing this case with one in which one part of my experience can be said to be an accompaniment of another. Thus we are sometimes tempted to ask: "If I feel this constant hum going on while I read, *where* is it?" I wish to make a pointing gesture, and there is nothing to point to. And the words "lay hold of" express the same misleading analogy.

Instead of asking the question "Where is this constant experience which seems to go on all through my reading?", we should ask "What is it in saying 'A particular atmosphere enshrouds the words which I am reading', that I am contrasting this case with?"

I will try to elucidate this by an analogous case: We are inclined to be puzzled by the three-dimensional appearance of the drawing

in a way expressed by the question "What does seeing it three-dimensionally consist in?" And this question really asks 'What is it that is added to simply seeing the drawing when we see it three-dimensionally?' And yet what answer can we expect to this question? It is the form of this question which produces the puzzlement. As Hertz says: "Aber offenbar irrt die Frage in Bezug auf die Antwort, welche sie erwartet" (p. 9, Einleitung, *Die Prinzipien der Mechanik*). The question itself keeps the mind pressing against a blank wall, thereby preventing it from ever finding the outlet. To show a man how to get out you have first of all to free him from the misleading influence of the question.

Look at a written word, say "read",—"It isn't just a scribble, it's 'read'", I should like to say, "it has one definite physiognomy". But what is it that I am really saying about it? What is this statement, straightened out? "The word falls", one is tempted to explain, "into a mould of my mind *long* prepared for it". But as I don't perceive both the word and a mould, the metaphor of the word's fitting a mould can't allude to an experience of comparing the hollow and the solid shape before they are fitted together, but rather to an experience of seeing the solid shape accentuated by a particular background.

i) would be the picture of the hollow and the solid shape before they are fitted together. We here see two circles and can compare them. *ii*) is the picture of the solid in the hollow. There is only one circle, and what we call the mould only accentuates, or as we sometimes said, emphasizes it.

I am tempted to say, "This isn't just a scribble, but it's *this* particular face".—But I can't say, "I see *this* as *this* face", but ought to say "I see this as *a* face". But I feel I want to say, "I don't see this as *a* face, I see it as *this* face". But in the second half of this sentence the word "face" is redundant, and it should have run, "I don't see this as a face, I see it like *this*".

Suppose I said "I see this scribble like *this*", and while saying "this scribble" I look at it as a mere scribble, and while saying "like *this*", I see the face,—this would come to something like saying "What at one time appears to me like this, at another appears to me like that", and here the "this" and the "that" would be accompanied by the two different ways of seeing.—But we must ask ourselves in what game is this sentence with the processes accompanying it to be used. E.g., to whom am I telling this? Suppose the answer is "I'm saying it to myself". But that is not enough. We are here in the grave danger of believing that we know what to do with a sentence if it looks more or less like one of the common sentences of our language. But here in order not to be deluded we have to ask ourselves: What is the use, say, of the words "this" and "that"?—or rather, What are the different uses which we make of them? What we call their meaning is not anything which they have got in them or which is fastened to them irrespective of what use we make of them. Thus it is one use of the word "this" to go along with a gesture pointing to something: We say "I am seeing the square with the diagonals like this", pointing to a swastika. And referring to the square with diagonals I might have said, "What at one time appears to me like this

at another time appears to me like that

."

And this is certainly not the use we made of the sentence in the above case.—One might think the whole difference between the two cases is this, that in the first the pictures are mental, in the second, real drawings. We should here ask ourselves in what sense we can call mental images pictures, for in some ways they are comparable to drawn or painted pictures, and in others not. It is, e.g., one of the essential points about the use of a 'material' picture that we say that it remains the same not only on the ground that it seems to us to be the same, that we remember that it looked before as it looks now. In fact we shall say under certain circumstances that the picture hasn't changed although it seems to have changed; and we say it hasn't changed because it has been kept in a certain way, certain influences have been kept out. Therefore the expression "The picture hasn't changed" is used in a different way when we talk of a material picture on the one hand, and of a mental one on the other. Just as the statement "These ticks follow at equal intervals" has got one grammar if the ticks are the tick of a pendulum and the criterion for their regularity is the result of measurements which we have made on our apparatus, and another grammar if the ticks are ticks which we imagine. I might for instance ask the question: When I said to myself "What at one time appears to me like this, at another . . .", did I recognize the two aspects, this and that, as the same which I got on previous occasions? Or were they new to me and I tried to remember them for future occasions? Or was all that I meant to say "I can change the aspect of this figure"?

19.   The danger of delusion which we are in becomes most clear if we propose to ourselves to give the aspects 'this' and 'that' names, say A and B. For we are most strongly tempted to imagine that giving a name consists in correlating in a peculiar and rather mysterious way a sound (or other sign) with something. How we make use of this peculiar correlation then seems to be almost a secondary matter. (One could almost imagine that naming was done by a peculiar sacramental act, and that this produced some magic relation between the name and the thing.)

But let us look at an example; consider this language game: A sends B to various houses in their town to fetch goods of various sorts from various people. A gives B various lists. On top of every list he puts a scribble, and B is trained to go to that house on the door of which he finds the same scribble; this is the name of the house. In the first column of every list he then finds one or more scribbles which he has been taught to read out. When he enters the house he calls out these words, and every inhabitant of the house has been trained to run up to him when a certain one of these sounds is called out, these sounds are the names of the people. He then addresses himself to each one of them in turn and shows to each two consecutive scribbles which stand on the list against his name. The first of these two, people of that town have been trained to associate with some particular kind of object, say, apples. The second is one of a series of scribbles which each man carries about him on a slip of paper. The person thus addressed fetches, say, five apples. The first scribble was the generic name of the objects required, the second, the name of their number.

What now is the relation between a name and the object named, say, the house and its name? I suppose we could give either of two answers. The one is that the relation consists in certain strokes having been painted on the door of the house. The second answer I meant is that the relation we are concerned with is established, not just by painting these strokes on the door, but by the particular role which they play in the practice of our language as we have been sketching it.—Again, the relation of the name of a person to the person here consists in the person having been trained to run up to someone who calls out the name; or again, we might say that it consists in this and the whole of the usage of the name in the language game.

Look into this language game and see if you can find the mysterious relation of the object and its name.—The relation of name and object we may say, consists in a scribble being written on an object (or some other such very trivial relation), and that's all there is to it. But we are not satisfied with that, for we feel that a scribble written on an object in itself is of no importance to us, and interests us in no way. And this is true; the whole importance lies in the particular use we make of the scribble written on the object, and we, in a sense, simplify matters by saying that the name has a peculiar relation to its object, a relation other than that, say, of being written on the object, or of being spoken by a person pointing to an object with his finger. A primitive philosophy condenses the whole usage of the name into the idea of a relation, which thereby becomes a mysterious relation. (Compare the ideas of mental activities, wishing, believing, thinking, etc., which for the same reason have something mysterious and inexplicable about them.)

Now we might use the expression "The relation of name and object does not merely consist in this kind of trivial, 'purely external', connection", meaning that what we call the relation of name and object is characterized by the entire usage of the name; but then it is clear that there is no one relation of name to object, but as many as there are uses of sounds or scribbles which we call names.

We can therefore say that if naming something is to be more than just uttering a sound while pointing to something, there must also be, in some form or other, the knowledge of how in the particular case the sound or scratch is to be used.

Now when we proposed to give the aspects of a drawing names, we made it appear that by seeing the drawing in two different ways, and each time saying something, we had done more than performing just this uninteresting action; whereas we now see that it is the usage of the 'name' and in fact the detail of this usage which gives the naming its peculiar significance.

It is therefore not an unimportant question, but a question about the essence of the matter; "Are 'A' and 'B' to remind me of these aspects; can I carry out such an order as 'See this drawing in the aspect A'; are there, in some way, pictures of these aspects correlated with the names 'A' and 'B' (like

and);

are 'A' and 'B' used in communicating with other people, and what exactly is the game played with them?"

When I say "I don't see mere dashes (a mere scribble) but a face (or word) with this particular physiognomy", I don't wish to assert any general characteristic of what I see, but to assert that I see that particular physiognomy which I do see. And it is obvious that here my expression is moving in a circle. But this is so because really the particular physiognomy which I saw ought to have entered my proposition.—When I find that "In reading a sentence, a peculiar experience goes on all the while", I have actually to read over a fairly long stretch to get the peculiar impression which makes one say this.

I might then have said "I find that the same experience goes on all the time", but I wished to say: "I don't just notice that it's the same experience throughout, I notice a particular experience". Looking at a uniformly coloured wall I might say, "I don't just see that it has the same colour all over, but I see a particular colour". But in saying this I am mistaking the function of a sentence.—It seems that you wish to specify the colour you see, but not by saying anything about it, nor by comparing it with a sample,—but by pointing to it; using it at the same time as the sample and that which the sample is compared with.

Consider this example: You tell me to write a few lines, and while I am doing so you ask "Do you feel something in your hand while you are writing?" I say, "Yes, I have a peculiar feeling".—Can't I say to myself when I write, "I have *this* feeling"? Of course I can say it, and while saying "this feeling", I concentrate on the feeling.—But what do I do with this sentence? What use is it to me? It seems that I am pointing out to myself what I am feeling,—as though my act of concentration was an 'inward' act of pointing, one which no one else but me is aware of, this however is unimportant. But I don't point to the feeling by attending to it. Rather, attending to the feeling means producing or modifying it. (On the other hand, observing a chair does not mean producing or modifying the chair.)

Our sentence "I have *this* feeling while I'm writing" is of the kind of the sentence "I see this". I don't mean the sentence when it is used to inform someone that I am looking at the object which I am pointing to, nor when it is used, as above, to convey to someone that I see a certain drawing in the way A and not in the way B. I mean the sentence, "I see this", as it is sometimes contemplated by us when we are brooding over certain philosophical problems. We are then, say, holding on to a particular visual impression by staring at some object, and we feel it is most natural to say to ourselves "I see this", though we know of no further use we can make of this sentence.

20. 'Surely it makes sense to say what I see, and how better could I do this than by letting what I see speak for itself!'

But the words "I see" in our sentence are redundant. I don't wish to tell myself that it is *I* who see this, nor that I *see* it. Or, as I might put it, it is impossible that I should not see *this*. This comes to the same as saying that I can't point out to myself by a visual hand what I am seeing; as this hand does not point to what I see but is part of what I see.

It is as though the sentence was singling out the particular colour I saw; as if it presented it to me.

It seems as though the colour which I see was its own description.

For the pointing with my finger was ineffectual. (And the looking is no pointing, it does not, for me, indicate a direction, which would mean contrasting a direction with other directions.)

What I see, or feel, enters my sentence as a sample does; but no use is made of this sample; the words of my sentence don't seem to matter, they only serve to present the sample to me.

I don't really speak *about* what I see, but *to* it.

I am in fact going through the acts of attending which could accompany the use of a sample. And this is what makes it seem as though I was making use of a sample. This error is akin to that of believing that an ostensive definition says something about the object to which it directs our attention.

When I said "I am mistaking the function of a sentence" it was because by its help I seemed to be pointing out to myself which colour it is I see, whereas I was just contemplating a sample of a colour. It seemed to me that the sample was the description of its own colour.

21.   Suppose I said to someone: "Observe the particular lighting of this room".—Under certain circumstances the sense of this order will be quite clear, e.g., if the walls of the room were red with the setting sun. But suppose at any other time when there is nothing striking about the lighting I said "Observe the particular lighting of this room":—Well, isn't there a particular lighting? So what is the difficulty about observing it? But the person who was told to observe the lighting when there was nothing striking about it would probably look about the room and say "Well, what about it?" Now I might go on and say "It is exactly the same lighting as yesterday at this hour", or "It is just this slightly dim light which you see in this picture of the room".

In the first case, when the room was lit a striking red, you could have pointed out the peculiarity which you were meant, though not explicitly told, to observe. You could, e.g., have used a sample of the particular colour in order to do so. We shall in this case be inclined to say that a peculiarity was added to the normal appearance of the room.

In the second case, when the room was just ordinarily lighted and there was nothing striking about its appearance, you didn't know exactly what to do when you were told to observe the lighting of the room. All you could do was to look about you waiting for something further to be said which would give the first order its full sense.

But wasn't the room, in both cases, lit in a particular way? Well, this question, as it stands, is senseless, and so is the answer "It was . . .". The order "Observe the particular lighting of this room" does not imply any statement about the appearance of this room. It seemed to say: "This room has a particular lighting, which I need not name; observe it!" The lighting referred to, it seems, is given by a sample, and you are to make use of the sample; as you would be doing in copying the precise shade of a colour sample on a palette. Whereas the order is similar to this: "Get hold of this sample!"

Imagine yourself saying "There is a particular lighting which I'm to observe". You could imagine yourself in this case staring about you in vain, that is, without seeing the lighting.

You could have been given a sample, e.g., a piece of colour material, and been asked: "Observe the colour of this patch".— And we can draw a distinction between observing, attending to, the shape of the sample and attending to its colour. But, attending to the colour can't be described as looking at a thing which is connected with the sample, rather, as looking at the sample in a peculiar way.

When we obey the order, "Observe the colour . . .", what we do is to open our eyes to colour. "Observe the colour . . ." doesn't mean "See the colour you see". The order, "Look at so and so", is of the kind, "Turn your head in this direction"; what you will see when you do so does not enter this order. By attending, looking, you produce the impression; you can't look at the impression.

Suppose someone answered to our order: "All right, I am now observing the particular lighting this room has",—this would sound as though he could point out to us which lighting it was. The order, that is to say, may seem to have told you to do something with this particular lighting, as opposed to another one (like "Paint this lighting, not that"). Whereas you obey the order by taking in *lighting*, as opposed to dimensions, shapes, etc.

(Compare, "Get hold of the colour of this sample" with "Get hold of this pencil", i.e., there it is, take hold of it.)

I return to our sentence: "this face has a particular expression". In this case too I did not compare or contrast my impression with anything, I did not make use of the sample before me. The sentence was an utterance* of a state of attention.

What has to be explained is this: Why do we talk to our impression?—You read, put yourself into a state of attention and say: "Something peculiar happens undoubtedly". You are inclined to go on: "There is a certain smoothness about it"; but you feel that this is only an inadequate description and that the experience can only stand for itself. "Something peculiar happens undoubtedly" is like saying, "I have had an experience". But you don't wish to make a general statement independent of the particular experience you have had, but rather a statement into which this experience enters.

You are under an impression. This makes you say "I am under a *particular* impression", and this sentence seems to say, to yourself at least, under what impression you are. As though you were referring to a picture ready in your mind, and said "This is what my impression is like". Whereas you have only pointed to your impression. In our case (p. 300), saying "I notice the particular colour of this wall" is like drawing, say, a black rectangle enclosing a small patch of the wall and thereby designating that patch as a sample for further use.

When you read, as it were attending closely to what happened in reading, you seemed to be observing reading as under a magnifying glass and to see the reading process. (But the case is more like that of observing something through a coloured glass.) You

---

* I.e. Außerung. See *Philosophical Investigations*, §256.—*Edd.*

think you have noticed the process of reading, the particular way in which signs are translated into spoken words.

22. I have read a line with a peculiar attention; I am impressed by the reading, and this makes me say that I have observed something besides the mere seeing of the written signs and the speaking of words. I have also expressed it by saying that I have noticed a particular atmosphere round the seeing and speaking. How such a metaphor as that embodied in the last sentence can come to suggest itself to me may be seen more clearly by looking at this example: If you heard sentences spoken in a monotone, you might be tempted to say that the words were all enshrouded in a particular atmosphere. But wouldn't it be using a peculiar way of representation to say that speaking the sentence in a monotone was adding something to the mere saying of it? Couldn't we even conceive speaking in a monotone as the result of *taking away* from the sentence its inflexion? Different circumstances would make us adopt different ways of representation. If, e.g., certain words had to be read out in a monotone, this being indicated by a staff and a sustained note beneath the written words, this notation would very strongly suggest the idea that something had been added to the mere speaking of the sentence.

I am impressed by the reading of a sentence, and I say the sentence has shown me something, that I have noticed something in it. This made me think of the following example: A friend and I once looked at beds of pansies. Each bed showed a different kind. We were impressed by each in turn. Speaking about them my friend said "What a variety of colour patterns, and each says something". And this was just what I myself wished to say.

Compare such a statement with this: "Every one of these men says something".—

If one had asked what the colour pattern of the pansy said, the right answer would have seemed to be that it said itself.

Hence we could have used an intransitive form of expression, say "Each of these colour patterns impresses one".

It has sometimes been said that what music conveys to us are feelings of joyfulness, melancholy, triumph, etc., etc. and what repels us in this account is that it seems to say that music is an instrument for producing in us sequences of feelings. And from this one might gather that any other means of producing such feelings would do for us instead of music.—To such an account we are tempted to reply "Music conveys to us *itself*!"

It is similar with such expressions as "Each of these colour patterns impresses one". We feel we wish to guard against the idea that a colour pattern is a means to producing in us a certain impression—the colour pattern being like a drug and we interested merely in the effect this drug produces.—We wish to avoid any form of expression which would seem to refer to an effect produced by an object on a subject. (Here we are bordering on the problem of idealism and realism and on the problem whether statements of aesthetics are subjective or objective.) Saying, "I see this and am impressed" is apt to make it seem as though the impression was some feeling accompanying the seeing, and that the sentence said something like "I see this and feel a pressure".

I could have used the expression "Each of these colour patterns has meaning"; but I didn't say "has meaning", for this would provoke the question, "What meaning?", which in the case we are considering is senseless. We are distinguishing between meaningless patterns and patterns which have meaning; but there is no such expression in our game as "This pattern has the meaning so and so". Nor even the expression "These two patterns have different meanings", unless this is to say: "These are two different patterns and both have meaning".

It is easy to understand though why we should be inclined to use the transitive form of expression. For let us see what use we make of such an expression as "This face says something", that is, what the situations are in which we use this expression, what sentences would precede or follow it (what kind of conversation it is a part of). We should perhaps follow up such a remark by saying, "Look at the line of these eyebrows" or "The *dark* eyes and the *pale* face!", these expressions would draw attention to certain features. We should in the same connection use comparisons, as for instance, "The nose is like a beak",—but also such expressions as "The whole face expresses bewilderment", and here we have used "expressing" transitively.

23.   We can now consider sentences which, as one might say, give an analysis of the impression we get, say, from a face. Take such a statement as, "The particular impression of this face is due to its small eyes and low forehead". Here the words "the particular impression" may stand for a certain specification, e.g., "the stupid expression". Or, on the other hand, they may mean 'what makes this expression a striking one' (i.e., an extraordinary one); or, 'what strikes one about this face' (i.e., 'what draws one's attention'). Or again, our sentence may mean "If you change *these* features in the slightest the expression will change entirely (whereas you might change other features without changing the expression nearly so much)". The form of this statement, however, mustn't mislead us into thinking that there is in every case a supplementing statement of the form "First the expression was *this*, after the change it's *that*". We can, of course, say "Smith frowned, and his expression changed from this to that", pointing, say, at two drawings of his face.—(Compare with this the two statements: "He said these words", and "His words said something".)

When, trying to see what reading consisted in, I read a written sentence, let the reading of it impress itself upon me, and said that I had a particular impression, one could have asked me such a question as whether my impression was not due to the particular handwriting. This would be asking me whether my impression would not be a different one if the writing had been a different one, or say, if each word of the sentence were written in a different handwriting. In this sense we could also ask whether that impression wasn't due after all to the *sense* of the particular sentence which I read. One might suggest: Read a different sentence (or the same one in a different handwriting) and see if you would still say that you had the same impression. And the answer might be: "Yes, the impression I had was really due to the handwriting".—But this would *not* imply that when I first said the sentence gave me a particular impression I had contrasted one impression with another, or that my statement had not been of the kind "This sentence has *its own character*". This will get clearer by considering the following example: Suppose we have three faces drawn side by side:

I contemplate the first one, saying to myself "This face has a peculiar expression". Then I am shown the second one and asked whether it has the same expression. I answer "Yes". Then the third one is shown to me and I say "It has a different expression". In my two answers I might be said to have distinguished the face and its expression: for b) is different from a) and still I say they have the same expression, whereas the difference between c) and a) corresponds to a difference of expression; and this may make us think that also in my first utterance I distinguished between the face and its expression.

24.   Let us now go back to the idea of a feeling of familiarity, which arises when I see familiar objects. Pondering about the question whether there is such a feeling or not, we are likely to gaze at some object and say, "Don't I have a particular feeling when I look at my old coat and hat?" But to this we now answer: What feeling do you compare this with, or oppose it to? Should

you say that your old coat gives you the same feeling as your old friend A with whose appearance too you are well acquainted, or that *whenever* you happened to look at your coat you get that feeling, say of intimacy and warmth?

'But is there no such thing as a feeling of familiarity?'—I should say that there are a great many different experiences, some of them feelings, which we might call "experiences (feelings) of familiarity".

Different experiences of familiarity: *a*) Someone enters my room, I haven't seen him for a long time, and didn't expect him. I look at him, say or feel "Oh, it's you".—Why did I in giving this example say that I hadn't seen the man for a long time? Wasn't I setting out to describe *experiences of* familiarity? And whatever the experience was I alluded to, couldn't I have had it even if I had seen the man half an hour ago? I mean, I gave the circumstances of recognizing the man as a means to the end of describing the precise situation of the recognition. One might object to this way of describing the *experience*, saying that it brought in irrelevant things, and in fact wasn't a *description* of the feeling at all. In saying this one takes as the prototype of a description, say, the description of a table, which tells you the exact shape, dimensions, the material which it is made of, and its colour. Such a description one might say pieces the table together. There is on the other hand a different kind of description of a table, such as you might find in a novel, e.g., "It was a small rickety table decorated in Moorish style, the sort that is used for smoker's requisites". Such a description might be called an indirect one; but if the purpose of it is to bring a vivid image of the table before your mind in a flash, it might serve this purpose incomparably better than a detailed 'direct' description.—Now if I am to give the description of a feeling of familiarity or recognition,—what do you expect me to do? Can I piece the feeling together? In a sense of course I could, giving you many different stages and the way my feelings changed. Such detailed descriptions you can find in some of the great novels. Now if you think of descriptions of pieces of furniture as you might find them in a novel, you see that to this kind of description you can oppose another making use of drawings, measures such as one should give to a cabinet maker. This latter kind one is inclined to call the only direct and complete description (though this way of expressing ourselves shows that we forget that there are certain purposes which the 'real' description does not fulfil). These considerations should warn you not to think that there is one real and direct description of, say, the feeling of recognition as opposed to the 'indirect' one which I have given.

b) The same as *a*), but the face is not familiar to me immediately. After a little, recognition 'dawns upon me'. I say, "Oh, it's you", but with totally different inflexion than in *a*). (Consider tone of voice, inflexion, gestures, as *essential* parts of our experience, not as inessential accompaniments or mere means of communication. (Compare pp. 263–66.)) *c*) There is an experience directed towards people or things which we see every day when suddenly we feel them to be 'old acquaintances' or 'good old friends'; one might also describe the feeling as one of warmth or of being at home with them. *d*) My room with all the objects in it is thoroughly familiar to me. When I enter it in the morning do I greet the familiar chairs, tables, etc., with a feeling of "Oh, hello!"? or have such a feeling as described in *c*)? But isn't the way I walk about in it, take something out of a drawer, sit down, etc., different from my behaviour in a room I don't know? And why shouldn't I say therefore, that I had experiences of familiarity whenever I lived amongst these familiar objects? *e*) Isn't it an experience of familiarity when on being asked "Who is this man?" I answer straight away (or after some reflection) "It is so and so"? Compare with this experience, *f*), that of looking at the written word "feeling" and saying "This is A's handwriting" and on the other hand *g*) the experience of reading the word, which also is an experience of familiarity.

To *e*) one might object, saying that the experience of saying the man's name was not the experience of familiarity, that he had to be familiar to us in order that we might know his name, and that we had to *know his name* in order that we might say it. Or, we might say "Saying his name is not enough, for surely we might say the name without knowing that it was his name". And this remark is certainly true if only we realize that it does not imply that knowing the name is a process accompanying or preceding saying the name.

25.   Consider this example: What is the difference between a memory image, an image that comes with expectation, and say, an image of a daydream. You may be inclined to answer, "There is an intrinsic difference between the images".—Did you notice that difference, or did you only say there was one because you think there must be one?

But surely I recognize a memory image as a memory image, an image of a daydream as an image of a daydream, etc.!— Remember that you are sometimes doubtful whether you actually saw a certain event happening or whether you dreamt it, or just had heard of it and imagined it vividly. But apart from that, what do you mean by "recognizing an image as a memory image"? I agree that (at least in most cases) while an image is before your mind's eye you are not in a state of doubt as to whether it is a memory image, etc. Also, if asked whether your image was a memory image, you would (in most cases) answer the question without hesitation. Now what if I asked you "*When* do you know what sort of an image it is?" Do you call knowing what sort of image it is not being in a state of doubt, not wondering about it? Does introspection make you see a state or activity of mind which you would call knowing that the image was a memory image, and which takes place while the image is before your mind?—Further, if you answer the question what sort of image it was you had, do you do so by, as it were, looking at the image and discovering a certain characteristic in it (as though you had been asked by whom a picture was painted, looked at it, recognized the style, and said it was a Rembrandt)?

It is easy, on the other hand, to point out experiences characteristic of remembering, expecting, etc., accompanying the images, and further differences in the immediate or more remote surrounding of them. Thus we certainly *say* different things in the different cases, e.g., "I remember his coming into my room", "I expect his coming into my room", "I imagine his coming into my room".—"But surely this can't be all the difference there is!" It isn't all: There are the three different games played with these three words surrounding these statements.

When challenged: do we *understand* the word "remember", etc.? is there really a difference between the cases besides the mere verbal one? our thoughts move in the immediate surroundings of the image we had or the expression we used. I have an image of dining in Hall with T. If asked whether this is a memory image, I say "Of course", and my thoughts begin to move on paths starting from this image. I remember who sat next to us, what the conversation was about, what I thought about it, what happened to T later on, etc., etc.

Imagine two different games both played with chess men on a chess board. The initial positions of both are alike. One of the games is always played with red and green pieces, the other with black and white. Two people are beginning to play, they have the chess board between them with the green and red pieces in position. Someone asks them "Do you know what game you're intending to play?" A player answers "Of course; we are playing no. 2". "What is the difference now between playing no. 2 and no. 1?"— "Well, there are red and green pieces on the board and not black and white ones, also we say that we are playing no. 2".—"But this couldn't be the only difference; don't you *understand* what 'no. 2' means and what game the red and green pieces stand for?" Here we are inclined to say "Certainly I do", and to prove this to ourselves we actually begin to move the pieces according to the rules of game no. 2. This is what I should call moving in the immediate surrounding of our initial position.

But isn't there also a peculiar feeling of pastness characteristic of images as memory images? There certainly are experiences which I should be inclined to call feelings of pastness, although not always when I remember something is one of these feelings present.—To get clear about the nature of these feelings it is again very useful to remember that there are gestures of pastness and inflexions of pastness which we can regard as representing the experiences of pastness.

I will examine one particular case, that of a feeling which I shall roughly describe by saying it is the feeling of 'long, long ago'. These words and the tone in which they are said are a gesture of pastness. But I will specify the experience which I mean still further by saying that it is that corresponding to a certain tune (Davids Bündler Tänze—"Wie aus weiter Ferne"). I'm imagining this tune played with the right expression and thus recorded, say, for a gramophone. Then this is the most elaborate and exact expression of a feeling of pastness which I can imagine.

Now should I say that hearing this tune played with this expression is in itself that particular experience of pastness, or should I say that hearing the tune causes the feeling of pastness to arise and that this feeling accompanies the tune? I.e., can I separate what I call this experience of pastness from the experience of hearing the tune? Or, can I separate an experience of pastness expressed by a gesture from the experience of making this gesture? Can I discover something, the essential feeling of pastness, which remains after abstracting all those experiences which we might call the experiences of expressing the feeling?

I am inclined to suggest to you to put the expression of our experience in place of the experience. 'But these two aren't the same'. This is certainly true, at least in the sense in which it is true to say that a railway train and a railway accident aren't the same thing. And yet there is a justification for talking as though the expression "the gesture 'long, long ago'", and the expression "the feeling 'long, long ago'" had the same meaning. Thus I could give the rules of chess in the following way: I have a chess board before me with a set of chess men on it. I give rules for moving these particular chess men (these particular pieces of wood) on this particular board. Can these rules be the rules of the game of chess? They can be converted into them by the usage of a single operator, such as the word "any". Or, the rules for my particular set may stand as they are and be made into rules of the game of chess by changing our standpoint towards them.

There is the idea that the feeling, say, of pastness, is an amorphous something in a place, the mind, and that this something is the cause or effect of what we call the expression of feeling. The expression of feeling then is an indirect way of transmitting the feeling. And people have often talked of a direct transmission of feeling which would obviate the external medium of communication.

Imagine that I tell you to mix a certain colour and I describe the colour by saying that it is that which you get if you let sulphuric acid react on copper. This might be called an indirect way of communicating the colour I meant. It is conceivable that the reaction of sulphuric acid on copper under certain circumstances does not produce the colour I wished you to mix, and that on seeing the colour you had got I should have to say "No, it's not this", and to give you a sample.

Now can we say that the communication of feelings by gestures is in this sense indirect? Does it make sense to talk of a direct communication as opposed to that indirect one? Does it make sense to say "I can't feel his toothache, but if I could I'd know what he feels like"?

If I speak of communicating a feeling to someone else, mustn't

ON CERTAINTY

Edited by G. E. M. Anscombe and G. H. von Wright

Translated by Denis Paul and G. E. M. Anscombe

1. If you do know that *here is one hand*,* we'll grant you all the rest.

   When one says that such and such a proposition can't be proved, of course that does not mean that it can't be derived from other propositions; any proposition can be derived from other ones. But they may be no more certain than it is itself. (On this a curious remark by H. Newman.)

2. From its *seeming* to me—or to everyone—to be so, it doesn't follow that it *is* so.

   What we can ask is whether it can make sense to doubt it.

---

* See G. E. Moore, "Proof of an External World", *Proceedings of the British Academy,* Vol. XXV, 1939; also "A Defence of Common Sense" in *Contemporary British Philosophy, 2nd Series,* Ed. J. H. Muirhead, 1925. Both papers are in Moore's *Philosophical Papers,* London, George Allen and Unwin, 1959. *Eds.*

w if there's a hand here"
—This possibility of satis-
guage-game. Is one of its

being." In order to see how
position is, consider its nega-
en to mean "I know I have the
brain which, after all, no one
t about such a proposition as "I
n I doubt it? Grounds for *doubt*
are lacking   speaks in its favour, nothing
against it. Nevertheless it is imaginable that my skull
should turn out empty when it was operated on.

5. Whether a proposition can turn out false after all depends
   on what I make count as determinants for that proposition.

6. Now, can one enumerate what one knows (like Moore)?
   Straight off like that, I believe not.—For otherwise the
   expression "I know" gets misused. And through this misuse a queer and extremely important mental state seems to
   be revealed.

7. My life shews that I know or am certain that there is a
   chair over there, or a door, and so on.—I tell a friend e.g.
   "Take that chair over there", "Shut the door", etc. etc.

8. The difference between the concept of 'knowing' and the
   concept of 'being certain' isn't of any great importance at
   all, except where "I know" is meant to mean: I *can't* be
   wrong. In a law-court, for example, "I am certain" could
   replace "I know" in every piece of testimony. We might
   even imagine its being forbidden to say "I know" there. [A

passage in *Wilhelm Meister*, where "You know" or "You knew" is used in the sense "You were certain", the facts being different from what he knew.]

9. Now do I, in the course of my life, make sure I know that here is a hand—my own hand, that is?

10. I know that a sick man is lying here? Nonsense! I am sitting at his bedside, I am looking attentively into his face.—So I don't know, then, that there is a sick man lying here? Neither the question nor the assertion makes sense. Any more than the assertion "I am here", which I might yet use at any moment, if suitable occasion presented itself.——Then is "2 × 2 = 4" nonsense in the same way, and not a proposition of arithmetic, apart from particular occasions? "2 × 2 = 4" is a true proposition of arithmetic—not "on particular occasions" nor "always"—but the spoken or written sentence "2 × 2 = 4" in Chinese might have a different meaning or be out and out nonsense, and from this is seen that it is only in use that the proposition has its sense. And "I know that there's a sick man lying here", used in an *unsuitable* situation, seems not to be nonsense but rather seems matter-of-course, only because one can fairly easily imagine a situation to fit it, and one thinks that the words "I know that . . ." are always in place where there is no doubt, and hence even where the expression of doubt would be unintelligible.

11. We just do not see how very specialized the use of "I know" is.

12. —For "I know" seems to describe a state of affairs which guarantees what is known, guarantees it as a fact. One always forgets the expression "I thought I knew".

13. For it is not as though the proposition "It is so" could be inferred from someone else's utterance: "I know it is so". Nor from the utterance together with its not being a lie.—But can't I infer "It is so" from my own utterance "I know etc."? Yes; and also "There is a hand there" follows from the proposition "He knows that there's a hand there". But from his utterance "I know . . ." it does not follow that he does know it.

14. That he does know takes some shewing.

15. It needs to be *shewn* that no mistake was possible. Giving the assurance "I know" doesn't suffice. For it is after all only an assurance that I can't be making a mistake, and it needs to be *objectively* established that I am not making a mistake about *that*.

16. "If I know something, then I also know that I know it, etc." amounts to: "I know that" means "I am incapable of being wrong about that". But whether I am so needs to be established objectively.

17. Suppose now I say "I'm incapable of being wrong about this: that is a book" while I point to an object. What would a mistake here be like? And have I any *clear* idea of it?

18. "I know" often means: I have the proper grounds for my statement. So if the other person is acquainted with the language-game, he would admit that I know. The other, if he is acquainted with the language-game, must be able to imagine *how* one may know something of the kind.

19. The statement "I know that here is a hand" may then be continued: "for it's *my* hand that I'm looking at". Then a

reasonable man will not doubt that I know.——Nor will the idealist; rather he will say that he was not dealing with the practical doubt which is being dismissed, but there is a further doubt *behind* that one.——That this is an *illusion* has to be shewn in a different way.

20. "Doubting the existence of the external world" does not mean for example doubting the existence of a planet, which later observations proved to exist.——Or does Moore want to say that knowing that here is his hand is different in kind from knowing the existence of the planet Saturn? Otherwise it would be possible to point out the discovery of the planet Saturn to the doubters and say that its existence has been proved, and hence the existence of the external world as well.

21. Moore's view really comes down to this: the concept 'know' is analogous to the concepts 'believe', 'surmise', 'doubt', 'be convinced' in that the statement "I know . . ." can't be a mistake. And if that *is* so, then there can be an inference from such an utterance to the truth of an assertion. And here the form "I thought I knew" is being overlooked.——But if this latter is inadmissible, then a mistake in the *assertion* must be logically impossible too. And anyone who is acquainted with the language-game must realize this—an assurance from a reliable man that he *knows* cannot contribute anything.

22. It would surely be remarkable if we had to believe the reliable person who says "I can't be wrong"; or who says "I am not wrong".

23. If I don't know whether someone has two hands (say, whether they have been amputated or not) I shall believe

his assurance that he has two hands, if he is trustworthy. And if he says he *knows* it, that can only signify to me that he has been able to make sure, and hence that his arms are e.g. not still concealed by coverings and bandages, etc. etc. My believing the trustworthy man stems from my admitting that it is possible for him to make sure. But someone who says that perhaps there are no physical objects makes no such admission.

24. The idealist's question would be something like: "What right have I not to doubt the existence of my hands?" (And to that the answer can't be: I *know* that they exist.) But someone who asks such a question is overlooking the fact that a doubt about existence only works in a language-game. Hence, that we should first have to ask: what would such a doubt be like?, and don't understand this straight off.

25. One may be wrong even about "there being a hand here". Only in particular circumstances is it impossible.—"Even in a calculation one can be wrong—only in certain circumstances one can't."

26. But can it be seen from a *rule* what circumstances logically exclude a mistake in the employment of rules of calculation?

   What use is a rule to us here? Mightn't we (in turn) go wrong in applying it?

27. If, however, one wanted to give something like a rule here, then it would contain the expression "in normal circumstances". And we recognize normal circumstances but cannot precisely describe them. At most, we can describe a range of abnormal ones.

28. What is 'learning a rule'?—*This*.

   What is 'making a mistake in applying it'?—*This*. And what is pointed to here is something indeterminate.

29. Practice in the use of the rule also shews what is a mistake in its employment.

30. When someone has made sure of something, he says: "Yes, the calculation is right", but he did not infer that from his condition of certainty. One does not infer how things are from one's own certainty.

   Certainty is *as it were* a tone of voice in which one declares how things are, but one does not infer from the tone of voice that one is justified.

31. The propositions which one comes back to again and again as if bewitched—these I should like to expunge from philosophical language.

32. It's not a matter of *Moore's* knowing that there's a hand there, but rather we should not understand him if he were to say "Of course I may be wrong about this". We should ask "What is it like to make such a mistake as that?"—e.g. what's it like to discover that it was a mistake?

33. Thus we expunge the sentences that don't get us any further.

34. If someone is taught to calculate, is he also taught that he can rely on a calculation of his teacher's? But these explanations must after all sometime come to an end. Will he also be taught that he can trust his senses—since he is indeed told in many cases that in such and such a special case you *cannot* trust them?—

   Rule and exception.

35. But can't it be imagined that there should be no physical objects? I don't know. And yet "There are physical objects" is nonsense. Is it supposed to be an empirical proposition?—

   And is *this* an empirical proposition: "There seem to be physical objects"?

36. "A is a physical object" is a piece of instruction which we give only to someone who doesn't yet understand either what "A" means, or what "physical object" means. Thus it is instruction about the use of words, and "physical object" is a logical concept. (Like colour, quantity, . . .) And that is why no such proposition as: "There are physical objects" can be formulated.

   Yet we encounter such unsuccessful shots at every turn.

37. But is it an adequate answer to the scepticism of the idealist, or the assurances of the realist, to say that "There are physical objects" is nonsense? For them after all it is not nonsense. It would, however, be an answer to say: this assertion, or its opposite is a misfiring attempt to express what can't be expressed like that. And that it does misfire can be shewn; but that isn't the end of the matter. We need to realize that what presents itself to us as the first expression of a difficulty, or of its solution, may as yet not be correctly expressed at all. Just as one who has a just censure of a picture to make will often at first offer the censure where it does not belong, and an *investigation* is needed in order to find the right point of attack for the critic.

38. Knowledge in mathematics: Here one has to keep on reminding oneself of the unimportance of the 'inner pro-

cess' or 'state' and ask "Why should it be important? What does it matter to me?" What is interesting is how we *use* mathematical propositions.

39. *This* is how calculation is done, in such circumstances a calculation is *treated* as absolutely reliable, as certainly correct.

40. Upon "I know that here is my hand" there may follow the question "How do you know?" and the answer to that presupposes that *this* can be known in *that* way. So, instead of "I know that here is my hand", one might say "Here is my hand", and then add *how* one knows.

41. "I know where I am feeling pain", "I know that I feel it *here*" is as wrong as "I know that I am in pain". But "I know where you touched my arm" is right.

42. One can say "He believes it, but it isn't so", but not "He knows it, but it isn't so". Does this stem from the difference between the mental states of belief and of knowledge? No.—One may for example call "mental state" what is expressed by tone of voice in speaking, by gestures etc. It would thus be *possible* to speak of a mental state of conviction, and that may be the same whether it is knowledge or false belief. To think that different states must correspond to the words "believe" and "know" would be as if one believed that different people had to correspond to the word "I" and the name "Ludwig", because the concepts are different.

43. What sort of proposition is this: "We *cannot* have miscalculated in $12 \times 12 = 144$"? It must surely be a proposition of logic.——But now, is it not the same, or doesn't it come to the same, as the statement $12 \times 12 = 144$?

44. If you demand a rule from which it follows that there can't have been a miscalculation here, the answer is that we did not learn this through a rule, but by learning to calculate.

45. We got to know the *nature* of calculating by learning to calculate.

46. But then can't it be described how we satisfy ourselves of the reliability of a calculation? O yes! Yet no rule emerges when we do so.—But the most important thing is: The rule is not needed. Nothing is lacking. We do calculate according to a rule, and that is enough.

47. *This* is how one calculates. Calculating is *this*. What we learn at school, for example. Forget this transcendent certainty, which is connected with your concept of spirit.

48. However, out of a host of calculations certain ones might be designated as reliable once for all, others as not yet fixed. And now, is this a *logical* distinction?

49. But remember: even when the calculation is something fixed for me, this is only a decision for a practical purpose.

50. When does one say, I know that . . . × . . . = . . . ? When one has checked the calculation.

51. What sort of proposition is: "What could a mistake here be like!"? It would have to be a logical proposition. But it is a logic that is not used, because what it tells us is not learned through propositions.—It is a logical proposition; for it does describe the conceptual (linguistic) situation.

52. This situation is thus not the same for a proposition like "At this distance from the sun there is a planet" and "Here is a hand" (namely my own hand). The second can't be called a hypothesis. But there isn't a sharp boundary line between them.

53. So one might grant that Moore was right, if he is interpreted like this: a proposition saying that here is a physical object may have the same logical status as one saying that here is a red patch.

54. For it is not true that a mistake merely gets more and more improbable as we pass from the planet to my own hand. No: at some point it has ceased to be conceivable.

    This is already suggested by the following: if it were not so, it would also be conceivable that we should be wrong in *every* statement about physical objects; that any we ever make are mistaken.

55. So is the *hypothesis* possible, that all the things around us don't exist? Would that not be like the hypothesis of our having miscalculated in all our calculations?

56. When one says: "Perhaps this planet doesn't exist and the light-phenomenon arises in some other way", then after all one needs an example of an object which does exist. This doesn't exist,—as *for example* does. . . .

    Or are we to say that *certainty* is merely a constructed point to which some things approximate more, some less closely? No. Doubt gradually loses its sense. This language-game just *is* like that.

    And everything descriptive of a language-game is part of logic.

57. Now might not "I *know*, I am not just surmising, that here is my hand" be conceived as a proposition of grammar? Hence *not* temporally.—

   But in that case isn't it like *this* one: "I know, I am not just surmising, that I am seeing red"?

   And isn't the consequence "So there are physical objects" like: "So there are colours"?

58. If "I know etc." is conceived as a grammatical proposition, of course the "I" cannot be important. And it properly means "There is no such thing as a doubt in this case" or "The expression 'I do not know' makes no sense in this case". And of course it follows from this that "I *know*" makes no sense either.

59. "I know" is here a *logical* insight. Only realism can't be proved by means of it.

60. It is wrong to say that the 'hypothesis' that *this* is a bit of paper would be confirmed or disconfirmed by later experience, and that, in "I know that this is a bit of paper," the "I know" either relates to such an hypothesis or to a logical determination.

61. . . . A meaning of a word is a kind of employment of it.

   For it is what we learn when the word is incorporated into our language.

62. That is why there exists a correspondence between the concepts 'rule' and 'meaning'.

63. If we imagine the facts otherwise than as they are, certain language-games lose some of their importance, while oth-

ers become important. And in this way there is an alteration—a gradual one—in the use of the vocabulary of a language.

64. Compare the meaning of a word with the 'function' of an official. And 'different meanings' with 'different functions'.

65. When language-games change, then there is a change in concepts, and with the concepts the meanings of words change.

---

66. I make assertions about reality, assertions which have different degrees of assurance. How does the degree of assurance come out? What consequences has it?

We may be dealing, for example, with the certainty of memory, or again of perception. I may be sure of something, but still know what test might convince me of error. I am e.g. quite sure of the date of a battle, but if I should find a different date in a recognized work of history, I should alter my opinion, and this would not mean I lost all faith in judging.

67. Could we imagine a man who keeps on making mistakes where we regard a mistake as ruled out, and in fact never encounter one?

E.g. he says he lives in such and such a place, is so and so old, comes from such and such a city, and he speaks with the same certainty (giving all the tokens of it) as I do, but he is wrong.

But what is his relation to this error? What am I to suppose?

68. The question is: what is the logician to say here?

69. I should like to say: "If I am wrong about *this*, I have no guarantee that anything I say is true." But others won't say that about me, nor will I say it about other people.

70. For months I have lived at address A, I have read the name of the street and the number of the house countless times, have received countless letters here and given countless people the address. If I am wrong about it, the mistake is hardly less than if I were (wrongly) to believe I was writing Chinese and not German.

71. If my friend were to imagine one day that he had been living for a long time past in such and such a place, etc. etc., I should not call this a *mistake*, but rather a mental disturbance, perhaps a transient one.

72. Not every false belief of this sort is a mistake.

73. But what is the difference between mistake and mental disturbance? Or what is the difference between my treating it as a mistake and my treating it as mental disturbance?

74. Can we say: a *mistake* doesn't only have a cause, it also has a ground? I.e., roughly: when someone makes a mistake, this can be fitted into what he knows aright.

75. Would this be correct: If I merely believed wrongly that there is a table here in front of me, this might still be a mistake; but if I believe wrongly that I have seen this table, or one like it, every day for several months past, and have regularly used it, that isn't a mistake?

76. Naturally, my aim must be to say what the statements one would like to make here, but cannot make significantly.

77. Perhaps I shall do a multiplication twice to make sure, or perhaps get someone else to work it over. But shall I work it over again twenty times, or get twenty people to go over it? And is that some sort of negligence? Would the certainty really be greater for being checked twenty times?

78. And can I give a *reason* why it isn't?

79. That I am a man and not a woman can be verified, but if I were to say I was a woman, and then tried to explain the error by saying I hadn't checked the statement, the explanation would not be accepted.

80. The *truth* of my statements is the test of my *understanding* of these statements.

81. That is to say: if I make certain false statements, it becomes uncertain whether I understand them.

82. What counts as an adequate test of a statement belongs to logic. It belongs to the description of the language-game.

83. The *truth* of certain empirical propositions belongs to our frame of reference.

84. Moore says he *knows* that the earth existed long before his birth. And put like that it seems to be a personal statement about him, even if it is in addition a statement about the physical world. Now it is philosophically uninteresting whether Moore knows this or that, but it is interesting that, and how, it can be known. If Moore had informed us that he knew the distance separating certain stars, we might conclude from that he had made some special investigations, and we shall want to know what these were. But

Moore chooses precisely a case in which we all seem to know the same as he, and without being able to say how. I believe e.g. that I know as much about this matter (the existence of the earth) as Moore does, and if he knows that it is as he says, then *I* know it too. For it isn't, either, as if he had arrived at his proposition by pursuing some line of thought which, while it is open to me, I have not in fact pursued.

85. And what goes into someone's knowing this? Knowledge of history, say? He must know what it means to say: the earth has already existed for such and such a length of time. For not *any* intelligent adult must know that. We see men building and demolishing houses, and are led to ask: "How long has this house been here?" But how does one come on the idea of asking this about a mountain, for example? And have all men the notion of the earth as a *body*, which may come into being and pass away? Why shouldn't I think of the earth as flat, but extending without end in every direction (including depth)? But in that case one might still say "I know that this mountain existed long before my birth."—But suppose I met a man who didn't believe that?

86. Suppose I replaced Moore's "I know" by "I am of the unshakeable conviction"?

87. Can't an assertoric sentence, which was capable of functioning as an hypothesis, also be used as a foundation for research and action? I.e. can't it simply be isolated from doubt, though not according to any explicit rule? It simply gets assumed as a truism, never called in question, perhaps not even ever formulated.

88. It may be for example that *all enquiry on our part* is set so as to exempt certain propositions from doubt, if they are ever formulated. They lie apart from the route travelled by enquiry.

89. One would like to say: "Everything speaks for, and nothing against the earth's having existed long before. . . ."

 Yet might I not believe the contrary after all? But the question is: What would the practical effects of this belief be?—Perhaps someone says: "That's not the point. A belief is what it is whether it has any practical effects or not." One thinks: It is the same adjustment of the human mind anyway.

90. "I know" has a primitive meaning similar to and related to "I see" ("wissen", "videre"). And "I knew he was in the room, but he wasn't in the room" is like "I saw him in the room, but he wasn't there". "I know" is supposed to express a relation, not between me and the sense of a proposition (like "I believe") but between me and a fact. So that the *fact* is taken into my consciousness. (Here is the reason why one wants to say that nothing that goes on in the outer world is really known, but only what happens in the domain of what are called sense-data.) This would give us a picture of knowing as the perception of an outer event through visual rays which project it as it is into the eye and the consciousness. Only then the question at once arises whether one can be *certain* of this projection. And this picture does indeed show how our *imagination* presents knowledge, but not what lies at the bottom of this presentation.

91. If Moore says he knows the earth existed etc., most of us will grant him that it has existed all that time, and also

believe him when he says he is convinced of it. But has he also got the right *ground* for his conviction? For if not, then after all he doesn't *know* (Russell).

92. However, we can ask: May someone have telling grounds for believing that the earth has only existed for a short time, say since his own birth?—Suppose he had always been told that,—would he have any good reason to doubt it? Men have believed that they could make rain; why should not a king be brought up in the belief that the world began with him? And if Moore and this king were to meet and discuss, could Moore really prove his belief to be the right one? I do not say that Moore could not convert the king to his view, but it would be a conversion of a special kind; the king would be brought to look at the world in a different way.

Remember that one is sometimes convinced of the *correctness* of a view by its *simplicity* or *symmetry*, i.e., these are what induce one to go over to this point of view. One then simply says something like: "*That's* how it must be."

93. The propositions presenting what Moore 'knows' are all of such a kind that it is difficult to imagine *why* anyone should believe the contrary. E.g. the proposition that Moore has spent his whole life in close proximity to the earth.—Once more I can speak of myself here instead of speaking of Moore. What could induce me to believe the opposite? Either a memory, or having been told.—Everything that I have seen or heard gives me the conviction that no man has ever been far from the earth. Nothing in my picture of the world speaks in favour of the opposite.

94. But I did not get my picture of the world by satisfying myself of its correctness; nor do I have it because I am

satisfied of its correctness. No: it is the inherited background against which I distinguish between true and
false.

95. The propositions describing this world-picture might be
part of a kind of mythology. And their role is like that of
rules of a game; and the game can be learned purely prac-
tically, without learning any explicit rules.

96. It might be imagined that some propositions, of the form
of empirical propositions, were hardened and functioned
as channels for such empirical propositions as were not
hardened but fluid; and that this relation altered with
time, in that fluid propositions hardened, and hard ones
became fluid.

97. The mythology may change back into a state of flux, the
river-bed of thoughts may shift. But I distinguish between
the movement of the waters on the river-bed and the shift
of the bed itself; though there is not a sharp division of the
one from the other.

98. But if someone were to say "So logic too is an empirical
science" he would be wrong. Yet this is right: the same
proposition may get treated at one time as something to
test by experience, at another as a rule of testing.

99. And the bank of that river consists partly of hard rock,
subject to no alteration or only to an imperceptible one,
partly of sand, which now in one place now in another
gets washed away, or deposited.

100. The truths which Moore says he knows, are such as,
roughly speaking, all of us know, if he knows them.

101. Such a proposition might be e.g. "My body has never disappeared and reappeared again after an interval."

102. Might I not believe that once, without knowing it, perhaps in a state of unconsciousness, I was taken far away from the earth—that other people even know this, but do not mention it to me? But this would not fit into the rest of my convictions at all. Not that I could describe the system of these convictions. Yet my convictions do form a system, a structure.

103. And now if I were to say "It is my unshakeable conviction that etc.", this means in the present case too that I have not consciously arrived at the conviction by following a particular line of thought, but that it is anchored in all my *questions and answers*, so anchored that I cannot touch it.

104. I am for example also convinced that the sun is not a hole in the vault of heaven.

105. All testing, all confirmation and disconfirmation of a hypothesis takes place already within a system. And this system is not a more or less arbitrary and doubtful point of departure for all our arguments: no, it belongs to the essence of what we call an argument. The system is not so much the point of departure, as the element in which arguments have their life.

106. Suppose some adult had told a child that he had been on the moon. The child tells me the story, and I say it was only a joke, the man hadn't been on the moon; no one has ever been on the moon; the moon is a long way off and it is impossible to climb up there or fly there.—If now the child insists, saying perhaps there is a way of getting there

which I don't know, etc. what reply could I make to him? What reply could I make to the adults of a tribe who believe that people sometimes go to the moon (perhaps that is how they interpret their dreams), and who indeed grant that there are no ordinary means of climbing up to it or flying there?—But a child will not ordinarily stick to such a belief and will soon be convinced by what we tell him seriously.

107. Isn't this altogether like the way one can instruct a child to believe in a God, or that none exists, and it will accordingly be able to produce apparently telling grounds for the one or the other?

108. "But is there then no objective truth? Isn't it true, or false, that someone has been on the moon?" If we are thinking within our system, then it is certain that no one has ever been on the moon. Not merely is nothing of the sort ever seriously reported to us by reasonable people, but our whole system of physics forbids us to believe it. For this demands answers to the questions "How did he overcome the force of gravity?" "How could he live without an atmosphere?" and a thousand others which could not be answered. But suppose that instead of all these answers we met the reply: "We don't know *how* one gets to the moon, but those who get there know at once that they are there; and even you can't explain everything." We should feel ourselves intellectually very distant from someone who said this.

109. "An empirical proposition can be *tested*" (we say). But how? and through what?

110. What *counts* as its test?—"But is this an adequate test? And, if so, must it not be recognizable as such in logic?"—As

if giving grounds did not come to an end sometime. But
the end is not an ungrounded presupposition: it is an ungrounded way of acting.

111. "I *know* that I have never been on the moon." That sounds
quite different in the circumstances which actually hold,
to the way it would sound if a good many men had been
on the moon, and some perhaps without knowing it. In
this case one could give grounds for this knowledge. Is
there not a relationship here similar to that between the
general rule of multiplying and particular multiplications
that have been carried out?

I want to say: my not having been on the moon is as
sure a thing for me as any grounds I could give for it.

112. And isn't that what Moore wants to say, when he says he
knows all these things?—But is his knowing it really what
is in question, and not rather that some of these propositions must be solid for us?

113. When someone is trying to teach us mathematics, he
will not begin by assuring us that he *knows* that
a+b = b+a.

114. If you are not certain of any fact, you cannot be certain of
the meaning of your words either.

115. If you tried to doubt everything you would not get as far
as doubting anything. The game of doubting itself presupposes certainty.

116. Instead of "I know . . .", couldn't Moore have said: "It
stands fast for me that . . ."? And further: "It stands fast
for me and many others. . . ."

117. Why is it not possible for me to doubt that I have never been on the moon? And how could I try to doubt it?

First and foremost, the supposition that perhaps I have been there would strike me as *idle*. Nothing would follow from it, nothing be explained by it. It would not tie in with anything in my life.

When I say "Nothing speaks for, everything against it," this presupposes a principle of speaking for and against. That is, I must be able to say what *would* speak for it.

118. Now would it be correct to say: So far no one has opened my skull in order to see whether there is a brain inside; but everything speaks for, and nothing against, its being what they would find there?

119. But can it also be said: Everything speaks for, and nothing against the table's still being there when no one sees it? For what does speak for it?

120. But if anyone were to doubt it, how would his doubt come out in practice? And couldn't we peacefully leave him to doubt it, since it makes no difference at all?

121. Can one say: "Where there is no doubt there is no knowledge either"?

122. Doesn't one need grounds for doubt?

123. Wherever I look, I find no ground for doubting that. . . .

124. I want to say: We use judgments as principles of judgment.

125. If a blind man were to ask me "Have you got two hands?" I should not make sure by looking. If I were to have any

doubt of it, then I don't know why I should trust my eyes. For why shouldn't I test my *eyes* by looking to find out whether I see my two hands? *What* is to be tested by *what*? (Who decides *what* stands fast?)

And what does it mean to say that such and such stands fast?

126. I am not more certain of the meaning of my words than I am of certain judgments. Can I doubt that this colour is called "blue"?

(My) doubts form a system.

127. For how do I know that someone is in doubt? How do I know that he uses the words "I doubt it" as I do?

128. From a child up I learned to judge like this. *This is* judging.

129. This is how I learned to judge; *this* I got to know *as* judgment.

130. But isn't it experience that teaches us to judge like *this*, that is to say, that it is correct to judge like this? But how does experience *teach* us, then? *We* may derive it from experience, but experience does not direct us to derive anything from experience. If it is the *ground* of our judging like this, and not just the cause, still we do not have a ground for seeing this in turn as a ground.

131. No, experience is not the ground for our game of judging. Nor is its outstanding success.

132. Men have judged that a king can make rain; *we* say this contradicts all experience. Today they judge that aero-

planes and the radio etc. are means for the closer contact of peoples and the spread of culture.

133. Under ordinary circumstances I do not satisfy myself that I have two hands by seeing how it looks. *Why* not? Has experience shown it to be unnecessary? Or (again): Have we in some way learnt a universal law of induction, and do we trust it here too?—But why should we have learnt one *universal* law first, and not the special one straight away?

134. After putting a book in a drawer, I assume it is there, unless. . . . "Experience always proves me right. There is no well attested case of a book's (simply) disappearing." It has *often* happened that a book has never turned up again, although we thought we knew for certain where it was.—But experience does really teach that a book, say, does not vanish away. (E.g. gradually evaporate.) But is it this experience with books etc. that leads us to assume that such a book has not vanished away? Well, suppose we were to find that under particular novel circumstances books did vanish away.—Shouldn't we alter our assumption? Can one give the lie to the effect of experience on our system of assumption?

135. But do we not simply follow the principle that what has always happened will happen again (or something like it)? What does it mean to follow this principle? Do we really introduce it into our reasoning? Or is it merely the *natural law* which our inferring apparently follows? This latter it may be. It is not an item in our considerations.

136. When Moore says he *knows* such and such, he is really enumerating a lot of empirical propositions which we affirm

without special testing; propositions, that is, which have a peculiar logical role in the system of our empirical propositions.

137. Even if the most trustworthy of men assures me that he *knows* things are thus and so, this by itself cannot satisfy me that he does know. Only that he believes he knows. That is why Moore's assurance that he knows . . . does not interest us. The propositions, however, which Moore retails as examples of such known truths are indeed interesting. Not because anyone knows their truth, or believes he knows them, but because they all have a *similar* role in the system of our empirical judgments.

138. We don't, for example, arrive at any of them as a result of investigation.

There are e.g. historical investigations and investigations into the shape and also the age of the earth, but not into whether the earth has existed during the last hundred years. Of course many of us have information about this period from our parents and grandparents; but mayn't they be wrong?—"Nonsense!" one will say. "How should all these people be wrong?"—But is that an argument? Is it not simply the rejection of an idea? And perhaps the determination of a concept? For if I speak of a possible mistake here, this changes the role of "mistake" and "truth" in our lives.

139. Not only rules, but also examples are needed for establishing a practice. Our rules leave loop-holes open, and the practice has to speak for itself.

140. We do not learn the practice of making empirical judgments by learning rules: we are taught *judgments* and their connexion with other judgments. *A totality* of judgments is made plausible to us.

141. When we first begin to *believe* anything, what we believe is not a single proposition, it is a whole system of propositions. (Light dawns gradually over the whole.)

142. It is not single axioms that strike me as obvious, it is a system in which consequences and premises give one another *mutual* support.

143. I am told, for example, that someone climbed this mountain many years ago. Do I always enquire into the reliability of the teller of this story, and whether the mountain did exist years ago? A child learns there are reliable and unreliable informants much later than it learns facts which are told it. It doesn't learn *at all* that that mountain has existed for a long time: that is, the question whether it is so doesn't arise at all. It swallows this consequence down, so to speak, together with *what* it learns.

144. The child learns to believe a host of things. I.e. it learns to act according to these beliefs. Bit by bit there forms a system of what is believed, and in that system some things stand unshakeably fast and some are more or less liable to shift. What stands fast does so, not because it is intrinsically obvious or convincing; it is rather held fast by what lies around it.

145. One wants to say "*All* my experiences shew that it is so". But how do they do that? For that proposition to which they point itself belongs to a particular interpretation of them.

    "That I regard this proposition as certainly true also characterizes my interpretation of experience."

146. We form *the picture* of the earth as a ball floating free in space and not altering essentially in a hundred years.

I said "We form the *picture* etc." and this picture now helps us in the judgment of various situations.

I may indeed calculate the dimensions of a bridge, sometimes calculate that here things are more in favour of a bridge than a ferry, etc. etc.,—but somewhere I must begin with an assumption or a decision.

147. The picture of the earth as a ball is a *good* picture, it proves itself everywhere, it is also a simple picture—in short, we work with it without doubting it.

148. Why do I not satisfy myself that I have two feet when I want to get up from a chair? There is no why. I simply don't. This is how I act.

149. My judgments themselves characterize the way I judge, characterize the nature of judgment.

150. How does someone judge which is his right and which his left hand? How do I know that my judgment will agree with someone else's? How do I know that this colour is blue? If I don't trust *myself* here, why should I trust anyone else's judgment? Is there a why? Must I not begin to trust somewhere? That is to say: somewhere I must begin with not-doubting; and that is not, so to speak, hasty but excusable: it is part of judging.

151. I should like to say: Moore does not *know* what he asserts he knows, but it stands fast for him, as also for me; regarding it as absolutely solid is part of our *method* of doubt and enquiry.

152. I do not explicitly learn the propositions that stand fast for me. I can *discover* them subsequently like the axis around

which a body rotates. This axis is not fixed in the sense that anything holds it fast, but the movement around it determines its immobility.

153. No one ever taught me that my hands don't disappear when I am not paying attention to them. Nor can I be said to presuppose the truth of this proposition in my assertions etc. (as if they rested on it) while it only gets sense from the rest of our procedure of asserting.

154. There are cases such that, if someone gives signs of doubt where we do not doubt, we cannot confidently understand his signs as signs of doubt.

I.e.: if we are to understand his signs of doubt as such, he may give them only in particular cases and may not give them in others.

155. In certain circumstances a man cannot make a *mistake*. ("Can" is here used logically, and the proposition does not mean that a man cannot say anything false in those circumstances.) If Moore were to pronounce the opposite of those propositions which he declares certain, we should not just not share his opinion: we should regard him as demented.

156. In order to make a mistake, a man must already judge in conformity with mankind.

157. Suppose a man could not remember whether he had always had five fingers or two hands? Should we understand him? Could we be sure of understanding him?

158. Can I be making a mistake, for example, in thinking that the words of which this sentence is composed are English words whose meaning I know?

159. As children we learn facts; e.g., that every human being has a brain, and we take them on trust. I believe that there is an island, Australia, of such-and-such a shape, and so on and so on; I believe that I had great-grandparents, that the people who gave themselves out as my parents really were my parents, etc. This belief may never have been expressed; even the thought that it was so, never thought.

160. The child learns by believing the adult. Doubt comes *after* belief.

161. I learned an enormous amount and accepted it on human authority, and then I found some things confirmed or disconfirmed by my own experience.

162. In general I take as true what is found in text-books, of geography for example. Why? I say: All these facts have been confirmed a hundred times over. But how do I know that? What is my evidence for it? I have a world-picture. Is it true or false? Above all it is the substratum of all my enquiring and asserting. The propositions describing it are not all equally subject to testing.

163. Does anyone ever test whether this table remains in existence when no one is paying attention to it?

    We check the story of Napoleon, but not whether all the reports about him are based on sense-deception, forgery and the like. For whenever we test anything, we are already presupposing something that is not tested. Now am I to say that the experiment which perhaps I make in order to test the truth of a proposition presupposes the truth of the proposition that the apparatus I believe I see is really there (and the like)?

164. Doesn't testing come to an end?

165. One child might say to another: "I know that the earth is already hundreds of years old" and that would mean: I have learnt it.

166. The difficulty is to realize the groundlessness of our believing.

167. It is clear that our empirical propositions do not all have the same status, since one can lay down such a proposition and turn it from an empirical proposition into a norm of description.

Think of chemical investigations. Lavoisier makes experiments with substances in his laboratory and now he concludes that this and that takes place when there is burning. He does not say that it might happen otherwise another time. He has got hold of a definite world-picture—not of course one that he invented: he learned it as a child. I say world-picture and not hypothesis, because it is the matter-of-course foundation for his research and as such also goes unmentioned.

168. But now, what part is played by the presupposition that a substance A always reacts to a substance B in the same way, given the same circumstances? Or is that part of the definition of a substance?

169. One might think that there were propositions declaring that chemistry is *possible*. And these would be propositions of a natural science. For what should they be supported by, if not by experience?

170. I believe what people transmit to me in a certain manner. In this way I believe geographical, chemical, historical facts etc. That is how I *learn* the sciences. Of course learning is based on believing.

     If you have learnt that Mont Blanc is 4000 metres high, if you have looked it up on the map, you say you *know* it.

     And can it now be said: we accord credence in this way because it has proved to pay?

171. A principal ground for Moore to assume that he never was on the moon is that no one ever was on the moon or *could* come there; and this we believe on grounds of what we learn.

172. Perhaps someone says "There must be some basic principle on which we accord credence", but what can such a principle accomplish? Is it more than a natural law of 'taking for true'?

173. Is it maybe in my power what I believe? or what I unshakeably believe?

     I believe that there is a chair over there. Can't I be wrong? But, can I believe that I am wrong? Or can I so much as bring it under consideration?—And mightn't I also hold fast to my belief whatever I learned later on?! But is my belief then *grounded*?

174. I act with *complete* certainty. But this certainty is my own.

175. "I know it" I say to someone else; and here there is a justification. But there is none for my belief.

176. Instead of "I know it" one may say in some cases "That's how it is—rely upon it." In some cases, however "I learned it years and years ago"; and sometimes: "I am sure it is so."

177. What I know, I believe.

178. The wrong use made by Moore of the proposition "I know . . ." lies in his regarding it as an utterance as little subject to doubt as "I am in pain". And since from "I know it is so" there follows "It is so", then the latter can't be doubted either.

179. It would be correct to say: "I believe . . ." has subjective truth; but "I know . . ." not.

180. Or again "I believe . . ." is an 'expression', but not "I know . . .".

181. Suppose Moore had said "I swear . . ." instead of "I know . . .".

182. The more primitive idea is that the earth *never* had a beginning. No child has reason to ask himself how long the earth has existed, because all change takes place *on* it. If what is called the earth really came into existence at some time—which is hard enough to picture—then one naturally assumes the beginning as having been an inconceivably long time ago.

183. "It is certain that after the battle of Austerlitz Napoleon. . . . Well, in that case it's surely also certain that the earth existed then."

184. "It is certain that we didn't arrive on this planet from another one a hundred years ago." Well, it's as certain as such things *are*.

185. It would strike me as ridiculous to want to doubt the existence of Napoleon; but if someone doubted the existence of the earth 150 years ago, perhaps I should be more willing to listen, for now he is doubting our whole system of evidence. It does not strike me as if this system were more certain than a certainty within it.

186. "I might suppose that Napoleon never existed and is a fable, but not that the *earth* did not exist 150 years ago."

187. "Do you *know* that the earth existed then?"—"Of course I know that. I have it from someone who certainly knows all about it."

188. It strikes me as if someone who doubts the existence of the earth at that time is impugning the nature of all historical evidence. And I cannot say of this latter that it is definitely *correct*.

189. At some point one has to pass from explanation to mere description.

190. What we call historical evidence points to the existence of the earth a long time before my birth;—the opposite hypothesis has *nothing* on its side.

191. Well, if everything speaks for an hypothesis and nothing against it—is it then certainly true? One may designate it as such.—But does it certainly agree with reality, with the

facts?—With this question you are already going round in a circle.

192. To be sure there is justification; but justification comes to an end.

———————————

193. What does this mean: the truth of a proposition is *certain*?

194. With the word "certain" we express complete conviction, the total absence of doubt, and thereby we seek to convince other people. That is *subjective* certainty.

    But when is something objectively certain? When a mistake is not possible. But what kind of possibility is that? Mustn't mistake be *logically* excluded?

195. If I believe that I am sitting in my room when I am not, then I shall not be said to have *made a mistake.* But what is the essential difference between this case and a mistake?

196. Sure evidence is what we *accept* as sure, it is evidence that we go by in *acting* surely, acting without any doubt.

    What we call "a mistake" plays a quite special part in our language games, and so too does what we regard as certain evidence.

197. It would be nonsense to say that we regard something as sure evidence because it is certainly true.

198. Rather, we must first determine the role of deciding for or against a proposition.

199. The reason why the use of the expression "true or false" has something misleading about it is that it is like saying "it tallies with the facts or it doesn't", and the very thing that is in question is what "tallying" is here.

200. Really "The proposition is either true or false" only means that it must be possible to decide for or against it. But this does not say what the ground for such a decision is like.

201. Suppose someone were to ask: "Is it really right for us to rely on the evidence of our memory (or our senses) as we do?"

202. Moore's certain propositions almost declare that we have a right to rely upon this evidence.

203. [Everything* that we regard as evidence indicates that the earth already existed long before my birth. The contrary hypothesis has *nothing* to confirm it at all.

     If everything speaks *for* an hypothesis and nothing against it, is it objectively *certain*? One can *call* it that. But does it *necessarily* agree with the world of facts? At the very best it shows us what "agreement" means. We find it difficult to imagine it to be false, but also difficult to make use of it.]

     What does this agreement consist in, if not in the fact that what is evidence in these language games speaks for our proposition? (*Tractatus Logico-Philosophicus*)

204. Giving grounds, however, justifying the evidence, comes to an end;—but the end is not certain propositions' striking us immediately as true, i.e. it is not a kind of *seeing* on our part; it is our *acting*, which lies at the bottom of the language-game.

---

* Passage crossed out in MS. *Eds.*

205. If the true is what is grounded, then the ground is not *true*, nor yet false.

206. If someone asked us "but is that *true*?" we might say "yes" to him; and if he demanded grounds we might say "I can't give you any grounds, but if you learn more you too will think the same".

    If this didn't come about, that would mean that he couldn't for example learn history.

207. "Strange coincidence, that every man whose skull has been opened had a brain!"

208. I have a telephone conversation with New York. My friend tells me that his young trees have buds of such and such a kind. I am now convinced that his tree is. . . . Am I also convinced that the earth exists?

209. The existence of the earth is rather part of the whole *picture* which forms the starting-point of belief for me.

210. Does my telephone call to New York strengthen my conviction that the earth exists?

    Much seems to be fixed, and it is removed from the traffic. It is so to speak shunted onto an unused siding.

211. Now it gives our way of looking at things, and our researches, their form. Perhaps it was once disputed. But perhaps, for unthinkable ages, it has belonged to the *scaffolding* of our thoughts. (Every human being has parents.)

212. In certain circumstances, for example, we regard a calculation as sufficiently checked. What gives us a right to do so? Experience? May that not have deceived us? Somewhere

we must be finished with justification, and then there remains the proposition that *this* is how we calculate.

213. Our 'empirical propositions' do not form a homogeneous mass.

214. What prevents me from supposing that this table either vanishes or alters its shape and colour when no one is observing it, and then when someone looks at it again changes back to its old condition?—"But who is going to suppose such a thing!"—one would feel like saying.

215. Here we see that the idea of 'agreement with reality' does not have any clear application.

216. The proposition "It is written".

217. If someone supposed that *all* our calculations were uncertain and that we could rely on none of them (justifying himself by saying that mistakes are always possible) perhaps we would say he was crazy. But can we say he is in error? Does he not just react differently? We rely on calculations, he doesn't; we are sure, he isn't.

218. Can I believe for one moment that I have ever been in the stratosphere? No. So do I *know* the contrary, like Moore?

219. There cannot be any doubt about it for me as a reasonable person.—That's it.—

220. The reasonable man does *not have* certain doubts.

221. Can I be in doubt at *will*?

222. I cannot possibly doubt that I was never in the strato-sphere. Does that make me know it? Does it make it true?

223. For mightn't I be crazy and not doubting what I absolutely ought to doubt?

224. "I *know* that it never happened, for if it had happened I could not possibly have forgotten it."

But, supposing it *did* happen, then it just would have been the case that you had forgotten it. And how do you know that you could not possibly have forgotten it? Isn't that just from earlier experience?

225. What I hold fast to is not *one* proposition but a nest of propositions.

226. Can I give the supposition that I have ever been on the moon any serious consideration at all?

227. "*Is* that something that one can forget?!"

228. "In such circumstances, people do not say 'Perhaps we've all forgotten', and the like, but rather they assume that . . ."

229. Our talk gets its meaning from the rest of our proceedings.

230. We are asking ourselves: what do we do with a statement "I *know* . . ."? For it is not a question of mental processes or mental states.

And *that* is how one must decide whether something is knowledge or not.

231. If someone doubted whether the earth had existed a hundred years ago, I should not understand, for *this* reason: I would not know what such a person would still allow to be counted as evidence and what not.

232. "We could doubt every single one of these facts, but we could not doubt them *all*."

    Wouldn't it be more correct to say: "we do not doubt them *all*".

    Our not doubting them all is simply our manner of judging, and therefore of acting.

233. If a child asked me whether the earth was already there before my birth, I should answer him that the earth did not begin only with my birth, but that it existed long, long before. And I should have the feeling of saying something funny. Rather as if the child had asked if such and such a mountain were higher than a tall house that it had seen. In answering the question I should have to be imparting a picture of the world to the person who asked it.

    If I do answer the question with certainty, what gives me this certainty?

234. I believe that I have forebears, and that every human being has them. I believe that there are various cities, and, quite generally, in the main facts of geography and history. I believe that the earth is a body on whose surface we move and that it no more suddenly disappears or the like than any other solid body: this table, this house, this tree, etc. If I wanted to doubt the existence of the earth long before my birth, I should have to doubt all sorts of things that stand fast for me.

235. And that something stands fast for me is not grounded in my stupidity or credulity.

236. If someone said "The earth has not long been . . ." what would he be impugning? Do I know?

Would it have to be what is called a scientific belief? Might it not be a mystical one? Is there any absolute necessity for him to be contradicting historical facts? Or even geographical ones?

237. If I say "an hour ago this table didn't exist" I probably mean that it was only made later on.

If I say "this mountain didn't exist then", I presumably mean that it was only formed later on—perhaps by a volcano.

If I say "this mountain didn't exist half an hour ago", that is such a strange statement that it is not clear what I mean. Whether for example I mean something untrue but scientific. Perhaps you think that the statement that the mountain didn't exist then is quite clear, however one conceives the context. But suppose someone said "This mountain didn't exist a minute ago, but an exactly similar one did instead". Only the accustomed context allows what is meant to come through clearly.

238. I might therefore interrogate someone who said that the earth did not exist before his birth, in order to find out which of my convictions he was at odds with. And then it *might be* that he was contradicting my fundamental attitudes that were how it was, and if I should have to put up with it.

Similarly if he said he had at some time been on the moon.

239. I believe that every human being has two human parents;
but Catholics believe that Jesus only had a human mother.
And other people might believe that there are human beings with no parents, and give no credence to all the contrary evidence. Catholics believe as well that in certain
circumstances a wafer completely changes its nature, and
at the same time that all evidence proves the contrary. And
so if Moore said "I know that this is wine and not blood",
Catholics would contradict him.

240. What is the belief that all human beings have parents
based on? On experience. And how can I base this sure
belief on my experience? Well, I base it not only on the fact
that I have known the parents of certain people but on
everything that I have learnt about the sexual life of human beings and their anatomy and physiology: also on
what I have heard and seen of animals. But then is that
really a proof?

241. Isn't this an hypothesis, which, as I *believe*, is again and
again completely confirmed?

242. Mustn't we say at every turn: "I *believe* this with certainty"?

243. One says "I know" when one is ready to give compelling
grounds. "I know" relates to a possibility of demonstrating
the truth. Whether someone knows something can come
to light, assuming that he is convinced of it.

But if what he believes is of such a kind that the grounds
that he can give are no surer than his assertion, then he
cannot say that he knows what he believes.

244. If someone says "I have a body", he can be asked "Who is speaking here with this mouth?"

245. To whom does anyone say that he knows something? To himself, or to someone else. If he says it to himself, how is it distinguished from the assertion that he is *sure* that things are like that? There is no subjective sureness that I know something. The certainty is subjective, but not the knowledge. So if I say "I know that I have two hands", and that is not supposed to express just my subjective certainty, I must be able to satisfy myself that I am right. But I can't do that, for my having two hands is not less certain before I have looked at them than afterwards. But I could say: "That I have two hands is an irreversible belief." That would express the fact that I am not ready to let anything count as a disproof of this proposition.

246. "Here I have arrived at a foundation of all my beliefs." "This position I will *hold*!" But isn't that, precisely, only because I am completely *convinced* of it?—What is 'being completely convinced' like?

247. What would it be like to doubt now whether I have two hands? Why can't I imagine it at all? What would I believe if I didn't believe that? So far I have no system at all within which this doubt might exist.

248. I have arrived at the rock bottom of my convictions.

And one might almost say that these foundation-walls are carried by the whole house.

249. One gives oneself a false picture of *doubt*.

250. My having two hands is, in normal circumstances, as certain as anything that I could produce in evidence for it.

    That is why I am not in a position to take the sight of my hand as evidence for it.

251. Doesn't this mean: I shall proceed according to this belief unconditionally, and not let anything confuse me?

252. But it isn't just that *I* believe in this way that I have two hands, but that every reasonable person does.

253. At the foundation of well-founded belief lies belief that is not founded.

254. Any 'reasonable' person behaves like *this*.

255. Doubting has certain characteristic manifestations, but they are only characteristic of it in particular circumstances. If someone said that he doubted the existence of his hands, kept looking at them from all sides, tried to make sure it wasn't 'all done by mirrors', etc., we should not be sure whether we ought to call that doubting. We might describe his way of behaving as like the behaviour of doubt, but his game would not be ours.

256. On the other hand a language-game does change with time.

257. If someone said to me that he doubted whether he had a body I should take him to be a half-wit. But I shouldn't know what it would mean to try to convince him that he

had one. And if I had said something, and that had removed his doubt, I should not know how or why.

258. I do not know how the sentence "I have a body" is to be used.

That doesn't unconditionally apply to the proposition that I have always been on or near the surface of the earth.

259. Someone who doubted whether the earth had existed for 100 years might have a scientific, or on the other hand a philosophical, doubt.

260. I would like to reserve the expression "I know" for the cases in which it is used in normal linguistic exchange.

261. I cannot at present imagine a reasonable doubt as to the existence of the earth during the last 100 years.

262. I can imagine a man who had grown up in quite special circumstances and been taught that the earth came into being 50 years ago, and therefore believed this. We might instruct him: the earth has long . . . etc.—We should be trying to give him our picture of the world.

This would happen through a kind of *persuasion*.

263. The schoolboy *believes* his teachers and his schoolbooks.

264. I could imagine Moore being captured by a wild tribe, and their expressing the suspicion that he has come from somewhere between the earth and the moon. Moore tells them that he knows etc. but he can't give them the grounds for his certainty, because they have fantastic

ideas of human ability to fly and know nothing about physics. This would be an occasion for making that statement.

265. But what does it say, beyond "I have never been to such and such a place, and have compelling grounds for believing that"?

266. And here one would still have to say what are compelling grounds.

267. "I don't merely have the visual impression of a tree: I *know* that it is a tree".

268. "I know that this is a hand."—And what is a hand?—"Well, *this*, for example."

269. Am I more certain that I have never been on the moon than that I have never been in Bulgaria? Why am I so sure? Well, I know that I have never been anywhere in the neighbourhood—for example I have never been in the Balkans.

270. "I have compelling grounds for my certitude." These grounds make the certitude objective.

271. What is a telling ground for something is not anything *I* decide.

272. I know = I am familiar with it as a certainty.

273. But when does one say of something that it is certain?
For there can be dispute whether something *is* certain; I mean, when something is *objectively* certain.

There are countless general empirical propositions that count as certain for us.

274. One such is that if someone's arm is cut off it will not grow again. Another, if someone's head is cut off he is dead and will never live again.

Experience can be said to teach us these propositions. However, it does not teach us them in isolation: rather, it teaches us a host of interdependent propositions. If they were isolated I might perhaps doubt them, for I have no experience relating to them.

275. If experience is the ground of our certainty, then naturally it is past experience.

And it isn't for example just *my* experience, but other people's, that I get knowledge from.

Now one might say that it is experience again that leads us to give credence to others. But what experience makes me believe that the anatomy and physiology books don't contain what is false? Though it is true that this trust is *backed up* by my own experience.

276. We believe, so to speak, that this great building exists, and then we see, now here, now there, one or another small corner of it.

277. "I can't help believing. . . ."

278. "I am comfortable that that is how things are."

279. It is quite sure that motor cars don't grow out of the earth. We feel that if someone could believe the contrary he could believe *everything* that we say is untrue, and could question everything that we hold to be sure.

But how does this *one* belief hang together with all the rest? We should like to say that someone who could believe that does not accept our whole system of verification.

This system is something that a human being acquires by means of observation and instruction. I intentionally do not say "learns".

280. After he has seen this and this and heard that and that, he is not in a position to doubt whether. . . .

281. *I*, L. W., believe, am sure, that my friend hasn't sawdust in his body or in his head, even though I have no direct evidence of my senses to the contrary. I am sure, by reason of what has been said to me, of what I have read, and of my experience. To have doubts about it would seem to me madness—of course, this is also in agreement with other people; but *I* agree with them.

282. I cannot say that I have good grounds for the opinion that cats do not grow on trees or that I had a father and a mother.

If someone has doubts about it—how is that supposed to have come about? By his never, from the beginning, having believed that he had parents? But then, is that conceivable, unless he has been taught it?

283. For how can a child immediately doubt what it is taught? That could mean only that he was incapable of learning certain language games.

284. People have killed animals since the earliest times, used the fur, bones etc. etc. for various purposes; they have

counted definitely on finding similar parts in any similar beast.

They have always learnt from experience; and we can see from their actions that they believe certain things definitely, whether they express this belief or not. By this I naturally do not want to say that men *should* behave like this, but only that they do behave like this.

285. If someone is looking for something and perhaps roots around in a certain place, he shows that he believes that what he is looking for is there.

286. What we believe depends on what we learn. We all believe that it isn't possible to get to the moon; but there might be people who believe that that is possible and that it sometimes happens. We say: these people do not know a lot that we know. And, let them be never so sure of their belief—they are wrong and we know it.

If we compare our system of knowledge with theirs then theirs is evidently the poorer one by far.

23.9.50

287. The squirrel does not infer by induction that it is going to need stores next winter as well. And no more do we need a law of induction to justify our actions or our predictions.

288. I know, not just that the earth existed long before my birth, but also that it is a large body, that this has been established, that I and the rest of mankind have forebears, that there are books about all this, that such books don't lie, etc. etc. etc. And I know all this? I believe it. This body of knowledge has been handed on to me and I have no

grounds for doubting it, but, on the contrary, all sorts of confirmation.

And why shouldn't I say that I know all this? Isn't that what one does say?

But not only I know, or believe, all that, but the others do too. Or rather, I *believe* that they believe it.

289. I am firmly convinced that others believe, believe they know, that all that is in fact so.

290. I myself wrote in my book that children learn to understand a word in such and such a way. Do I know that, or do I believe it? Why in such a case do I write not "I believe etc." but simply the indicative sentence?

291. We know that the earth is round. We have definitively ascertained that it is round.

We shall stick to this opinion, unless our whole way of seeing nature changes. "How do you know that?"—I believe it.

292. Further experiments cannot *give the lie* to our earlier ones, at most they may change our whole way of looking at things.

293. Similarly with the sentence "water boils at 100°C."

294. This is how we acquire conviction, this is called "being rightly convinced".

295. So hasn't one, in this sense, a *proof* of the proposition? But that the same thing has happened again is not a proof of it; though we do say that it gives us a right to assume it.

296. This is what we *call* an "empirical foundation" for our assumptions.

297. For we learn, not just that such and such experiments had those and those results, but also the conclusion which is drawn. And of course there is nothing wrong in our doing so. For this inferred proposition is an instrument for a definite use.

298. 'We are quite sure of it' does not mean just that every single person is certain of it, but that we belong to a community which is bound together by science and education.

299. We are satisfied that the earth is round.*

_____

10.3.51

300. Not all corrections of our views are on the same level.

301. Supposing it wasn't true that the earth had already existed long before I was born—how should we imagine the mistake being discovered?

302. It's no good saying "Perhaps we are wrong" when, if *no* evidence is trustworthy, trust is excluded in the case of the present evidence.

303. If, for example, we have always been miscalculating, and twelve times twelve isn't a hundred and forty-four, why should we trust any other calculation? And of course that is wrongly put.

_____

* In English. *Eds.*

304. But nor am I *making a mistake* about twelve times twelve being a hundred and forty-four. I may say later that I was confused just now, but not that I was making a mistake.

305. Here *once more* there is needed a step like the one taken in relativity theory.

306. "I don't know if this is a hand." But do you know what the word "hand" means? And don't say "I know what it means now for me". And isn't it an empirical fact—that *this* word is used like *this*?

307. And here the strange thing is that when I am quite certain of how the words are used, have no doubt about it, I can still give no *grounds* for my way of going on. If I tried I could give a thousand, but none as certain as the very thing they were supposed to be grounds for.

308. 'Knowledge' and 'certainty' belong to different *categories*. They are not two 'mental states' like, say 'surmising' and 'being sure'. (Here I assume that it is meaningful for me to say "I know what [e.g.] the word 'doubt' means" and that this sentence indicates that the word "doubt" has a logical role.) What interests us now is not being sure but knowledge. That is, we are interested in the fact that about certain empirical propositions no doubt can exist if making judgments is to be possible at all. Or again: I am inclined to believe that not everything that has the form of an empirical proposition *is* one.

309. Is it that rule and empirical proposition merge into one another?

310. A pupil and a teacher. The pupil will not let anything be explained to him, for he continually interrupts with doubts, for instance as to the existence of things, the meaning of words, etc. The teacher says "Stop interrupting me and do as I tell you. So far your doubts don't make sense at all".

311. Or imagine that the boy questioned the truth of history (and everything that connects up with it)—and even whether the earth had existed at all a hundred years before.

312. Here it strikes me as if this doubt were hollow. But in that case—isn't *belief* in history hollow too? No; there is so much that this connects up with.

313. So is *that* what makes us believe a proposition? Well—the grammar of "believe" just does hang together with the grammar of the proposition believed.

314. Imagine that the schoolboy really did ask "and is there a table there even when I turn round, and even when *no one* is there to see it?" Is the teacher to reassure him—and say "of course there is!"?

     Perhaps the teacher will get a bit impatient, but think that the boy will grow out of asking such questions.

315. That is to say, the teacher will feel that this is not really a legitimate question at all.

     And it would be just the same if the pupil cast doubt on the uniformity of nature, that is to say on the justification of inductive arguments.—The teacher would feel that this was only holding them up, that this way the pupil would only get stuck and make no progress.—And he would be right. It would be as if someone were looking for some object in a room; he opens a drawer and doesn't see it there; then he closes it again, waits, and opens it once more to see if perhaps it isn't there now, and keeps on like that. He has not learned to look for things. And in the same way this pupil has not learned how to ask questions. He has not learned *the* game that we are trying to teach him.

316. And isn't it the same as if the pupil were to hold up his history lesson with doubts as to whether the earth really. . . . ?

317. This doubt isn't one of the doubts in our game. (But not as if we *chose* this game!)

12.3.51

318. 'The question doesn't arise at all.' Its answer would characterize a *method*. But there is no sharp boundary between methodological propositions and propositions within a method.

319. But wouldn't one have to say then, that there is no sharp boundary between propositions of logic and empirical propositions? The lack of sharpness *is* that of the boundary between *rule* and empirical proposition.

320. Here one must, I believe, remember that the concept 'proposition' itself is not a sharp one.

321. Isn't what I am saying: any empirical proposition can be transformed into a postulate—and then becomes a norm

of description. But I am suspicious even of this. The sentence is too general. One almost wants to say "any empirical proposition can, theoretically, be transformed . . .", but what does "theoretically" mean here? It sounds all too reminiscent of the *Tractatus*.

322. What if the pupil refused to believe that this mountain had been there beyond human memory?

We should say that he had no *grounds* for this suspicion.

323. So rational suspicion must have grounds?

We might also say: "the reasonable man believes this".

324. Thus we should not call anybody reasonable who believed something in despite of scientific evidence.

325. When we say that we *know* that such and such . . . , we mean that any reasonable person in our position would also know it, that it would be a piece of unreason to doubt it. Thus Moore too wants to say not merely that *he* knows that he etc. etc., but also that anyone endowed with reason in his position would know it just the same.

326. But who says what it is reasonable to believe in *this* situation?

327. So it might be said: "The reasonable man believes: that the earth has been there since long before his birth, that his life has been spent on the surface of the earth, or near it, that he has never, for example, been on the moon, that he has a nervous system and various innards like all other people, etc., etc."

328. "I know it *as* I know that my name is L. W."

329. 'If he calls *that* in doubt—whatever "doubt" means here—he will never learn this game'.

330. So here the sentence "I know . . ." expresses the readiness to believe certain things.

13.3.

331. If we ever do act with certainty on the strength of belief, should we wonder that there is much we cannot doubt?

332. Imagine that someone were to say, without wanting to *philosophize*, "I don't know if I have ever been on the moon; I don't *remember* ever having been there". (Why would this person be so radically different from us?)

   In the first place—how would he know that he was on the moon? How does he imagine it? Compare: "I do not know if I was ever in the village of X." But neither could I say that if X were in Turkey, for I know that I was never in Turkey.

333. I ask someone "Have you ever been in China?" He replies "I don't know". Here one would surely say "You don't *know*? Have you any reason to believe you might have been there at some time? Were you for example ever near the Chinese border? Or were your parents there at the time when you were going to be born?"—Normally Europeans do know whether they have been in China or not.

334. That is to say: only in such-and-such circumstances does a reasonable person doubt *that*.

335. The procedure in a court of law rests on the fact that circumstances give statements a certain probability. The
statement that, for example, someone came into the world
without parents wouldn't ever be taken into consideration
there.

336. But what men consider reasonable or unreasonable alters.
At certain periods men find reasonable what at other periods they found unreasonable. And vice versa.

But is there no objective character here?

Very intelligent and well-educated people believe in the
story of creation in the Bible, while others hold it as proven
false, and the grounds of the latter are well known to the
former.

337. One cannot make experiments if there are not some things
that one does not doubt. But that does not mean that one
takes certain presuppositions on trust. When I write a letter and post it, I take it for granted that it will arrive—I
expect this.

If I make an experiment I do not doubt the existence
of the apparatus before my eyes. I have plenty of doubts,
but not *that*. If I do a calculation I believe, without any
doubts, that the figures on the paper aren't switching of
their own accord, and I also trust my memory the whole
time, and trust it without any reservation. The certainty
here is the same as that of my never having been on the
moon.

338. But imagine people who were never quite certain of these
things, but said that they were *very* probably so, and that it
did not pay to doubt them. Such a person, then, would say
in my situation: "It is extremely unlikely that I have ever
been on the moon", etc., etc. *How* would the life of these

people differ from ours? For there *are* people who say that it is merely extremely probable that water over a fire will boil and not freeze, and that therefore strictly speaking what we consider impossible is only improbable. What difference does this make in their lives? Isn't it just that they talk rather more about certain things than the rest of us?

339. Imagine someone who is supposed to fetch a friend from the railway station and doesn't simply look the train up in the time-table and go to the station at the right time, but says: "I have *no* belief that the train will really arrive, but I will go to the station all the same." He does everything that the normal person does, but accompanies it with doubts or with self-annoyance, etc.

340. We know, with the same certainty with which we believe *any* mathematical proposition, how the letters A and B are pronounced, what the colour of human blood is called, that other human beings have blood and call it "blood".

341. That is to say, the *questions* that we raise and our *doubts* depend on the fact that some propositions are exempt from doubt, are as it were like hinges on which those turn.

342. That is to say, it belongs to the logic of our scientific investigations that certain things are *in deed* not doubted.

343. But it isn't that the situation is like this: We just *can't* investigate everything, and for that reason we are forced to rest content with assumption. If I want the door to turn, the hinges must stay put.

344. My *life* consists in my being content to accept many things.

345. If I ask someone "what colour do you see at the moment?", in order, that is, to learn what colour is there at the moment, I cannot at the same time question whether the person I ask understands English, whether he wants to take me in, whether my own memory is not leaving me in the lurch as to the names of colours, and so on.

346. When I am trying to mate someone in chess, I cannot have doubts about the pieces perhaps changing places of themselves and my memory simultaneously playing tricks on me so that I don't notice.

15.3.51

347. "I know that that's a tree." Why does it strike me as if I did not understand the sentence? though it is after all an extremely simple sentence of the most ordinary kind? It is as if I could not focus my mind on any meaning. Simply because I don't look for the focus where the meaning is. As soon as I think of an everyday use of the sentence instead of a philosophical one, its meaning becomes clear and ordinary.

348. Just as the words "I am here" have a meaning only in certain contexts, and not when I say them to someone who is sitting in front of me and sees me clearly,—and not because they are superfluous, but because their meaning is not *determined* by the situation, yet stands in need of such determination.

349. "I know that that's a tree"—this may mean all sorts of things: I look at a plant that I take for a young beech and that someone else thinks is a black-currant. He says "that

is a shrub"; I say it is a tree.—We see something in the mist which one of us takes for a man, and the other says "I know that that's a tree". Someone wants to test my eyes etc. etc.—etc. etc. Each time the 'that' which I declare to be a tree is of a different kind.

But what when we express ourselves more precisely? For example: "I know that that thing there is a tree, I can see it quite clearly."—Let us even suppose I had made this remark in the context of a conversation (so that it was relevant when I made it); and now, out of all context, I repeat it while looking at the tree, and I add "I mean these words as I did five minutes ago". If I added, for example, that I had been thinking of my bad eyes again and it was a kind of sigh, then there would be nothing puzzling about the remark.

For how a sentence is *meant* can be expressed by an expansion of it and may therefore be made part of it.

350. "I know that that's a tree" is something a philosopher might say to demonstrate to himself or to someone else that he *knows* something that is not a mathematical or logical truth. Similarly, someone who was entertaining the idea that he was no use any more might keep repeating to himself "I can still do this and this and this". If such thoughts often possessed him one would not be surprised if he, apparently out of all context, spoke such a sentence out loud. (But here I have already sketched a background, a surrounding, for this remark, that is to say given it a context.) But if someone, in quite heterogeneous circumstances, called out with the most convincing mimicry: "Down with him!", one might say of these words (and their tone) that they were a pattern that does indeed have familiar applications, but that in this case it was not even

clear what *language* the man in question was speaking. I might make with my hand the movement I should make if I were holding a hand-saw and sawing through a plank; but would one have any right to call this movement *sawing*, out of all context?—(It might be something quite different!)

351. Isn't the question "Have these words a meaning?" similar to "Is that a tool?" asked as one produces, say, a hammer? I say "Yes, it's a hammer". But what if the thing that any of us would take for a hammer were somewhere else a missile, for example, or a conductor's baton? Now make the application yourself.

352. If someone says, "I know that that's a tree" I may answer: "Yes, that is a sentence. An English sentence. And what is it supposed to be doing?" Suppose he replies: "I just wanted to remind myself that I *know* things like that"?———

353. But suppose he said "I want to make a logical observation"?———If a forester goes into a wood with his men and says "*This* tree has got to be cut down, and *this* one and *this* one"———what if he then observes "I *know* that that's a tree"?—But might not *I* say of the forester "He *knows* that that's a tree———he doesn't examine it, or order his men to examine it"?

354. Doubting and non-doubting behaviour. There is the first only if there is the second.

355. A mad-doctor (perhaps) might ask me "Do you know what that is?" and I might reply "I know that it's a chair; I

recognize it, it's always been in my room". He says this, possibly, to test not my eyes but my ability to recognize things, to know their names and their functions. What is in question here is a kind of knowing one's way about. Now it would be wrong for me to say "I believe that it's a chair" because that would express my readiness for my statement to be tested. While "I know that it . . ." implies *bewilderment* if what I said was not confirmed.

356. My 'mental state', the "knowing", gives me no guarantee of what will happen. But it consists in this, that I should not understand where a doubt could get a foothold nor where a further test was possible.

357. One might say: "'I know' expresses *comfortable* certainty, not the certainty that is still struggling."

358. Now I would like to regard this certainty, not as something akin to hastiness or superficiality, but as a form of life. (That is very badly expressed and probably badly thought as well.)

359. But that means I want to conceive it as something that lies beyond being justified or unjustified; as it were, as something animal.

360. I KNOW that this is my foot. I could not accept any experience as proof to the contrary.—That may be an exclamation; but what *follows* from it? At least that I shall act with a certainty that knows no doubt, in accordance with my belief.

361. But I might also say: It has been revealed to me by God that it is so. God has taught me that this is my foot. And therefore if anything happened that seemed to conflict

with this knowledge I should have to regard *that* as deception.

362. But doesn't it come out here that knowledge is related to a decision?

363. And here it is difficult to find the transition from the exclamation one would like to make, to its consequences in what one does.

364. One might also put this question: "If you know that that is your foot,—do you also know, or do you only believe, that no future experience will seem to contradict your knowledge?" (That is, that nothing will seem to *you yourself* to do so.)

365. If someone replied: "I also know that it will never *seem* to me as if anything contradicted that knowledge",—what could we gather from that, except that he himself had no doubt that it would never happen?—

366. Suppose it were forbidden to say "I know" and only allowed to say "I believe I know"?

367. Isn't it the purpose of construing a word like "know" analogously to "believe" that then opprobrium attaches to the statement "I know" if the person who makes it is wrong?

As a result a mistake becomes something forbidden.

368. If someone says that he will recognize no experience as proof of the opposite, that is after all a *decision*. It is possible that he will act against it.

16.3.51

369. If I wanted to doubt whether this was my hand, how could I avoid doubting whether the word "hand" has any meaning? So that is something I seem to *know* after all.

370. But more correctly: The fact that I use the word "hand" and all the other words in my sentence without a second thought, indeed that I should stand before the abyss if I wanted so much as to try doubting their meanings—shews that absence of doubt belongs to the essence of the language-game, that the question "How do I know . . ." drags out the language-game, or else does away with it.

371. Doesn't "I know that that's a hand", in Moore's sense, mean the same, or more or less the same, as: I can make statements like "I have a pain in this hand" or "this hand is weaker than the other" or "I once broke this hand", and countless others, in language-games where a doubt as to the existence of this hand does not come in?

372. Only in certain cases is it possible to make an investigation "is that really a hand?" (or "my hand"). For "I doubt whether that is really my (or a) hand" makes no sense without some more precise determination. One cannot tell from these words alone whether any doubt at all is meant—nor what kind of doubt.

373. Why should it be possible to have grounds for *believing* anything if it isn't possible to be certain?

374. We teach a child "that is your hand", not "that is perhaps [or "probably"] your hand". That is how a child learns the

innumerable language-games that are concerned with his hand. An investigation or question, 'whether this is really a hand' never occurs to him. Nor, on the other hand, does he learn that he *knows* that this is a hand.

375. Here one must realize that complete absence of doubt at some point, even where we would say that 'legitimate' doubt can exist, need not falsify a language-game. For there is also something like *another* arithmetic.

   I believe that this admission must underlie any understanding of logic.

17.3.

376. I may claim with passion that I know that this (for example) is my foot.

377. But this passion is after all something very rare, and there is no trace of it when I talk of this foot in the ordinary way.

378. Knowledge is in the end based on acknowledgment.

379. I say with passion "I *know* that this is a foot"—but what does it *mean*?

380. I might go on: "Nothing in the world will convince me of the opposite!" For me this fact is at the bottom of all knowledge. I shall give up other things but not this.

381. This "Nothing in the world" is obviously an attitude which one hasn't got towards everything one believes or is certain of.

382. That is not to say that nothing in the world will in fact be able to convince me of anything else.

383. The argument "I may be dreaming" is senseless for this reason: if I am dreaming, this remark is being dreamed as well—and indeed it is also being dreamed that these words have any meaning.

384. Now what kind of sentence is "Nothing in the world . . ."?

385. It has the form of a prediction, but of course it is not one that is based on experience.

386. Anyone who says, with Moore, that he knows that so and so . . .—gives the degree of certainty that something has for him. And it is important that this degree has a maximum value.

387. Someone might ask me: "How certain are you that that is a tree over there; that you have money in your pocket; that that is your foot?" And the answer in one case might be "not certain", in another "as good as certain", in the third "I can't doubt it". And these answers would make sense even without any grounds. I should not need, for example, to say: "I can't be certain whether that is a tree because my eyes aren't sharp enough". I want to say: it made sense for Moore to say "I *know* that that is a tree", if he meant something quite particular by it.

    [I believe it might interest a philosopher, one who can think himself, to read my notes. For even if I have hit the mark only rarely, he would recognize what targets I had been ceaselessly aiming at.]

388. Every one of us often uses such a sentence, and there is no question but that it makes sense. But does that mean it yields any philosophical conclusion? Is it more of a proof of the existence of external things, that I know that this is a hand, than that I don't know whether that is gold or brass?

18.3.

389. Moore wanted to give an example to shew that one really can *know* propositions about physical objects.—If there were a dispute whether one could have a pain in such and such a part of the body, then someone who just then had a pain in that spot might say: "I assure you, I have a pain there now." But it would sound odd if Moore had said: "I assure you, I know that's a tree." A personal experience simply has no interest for us here.

390. All that is important is that it makes sense to say that one knows such a thing; and consequently the assurance that one does know it can't accomplish anything here.

391. Imagine a language-game "When I call you, come in through the door". In any ordinary case, a doubt whether there really is a door there will be impossible.

392. What I need to shew is that a doubt is not necessary even when it is possible. That the possibility of the language-game doesn't depend on everything being doubted that can be doubted. (This is connected with the role of contradiction in mathematics.)

393. The sentence "I know that that's a tree" if it were said outside its language-game, might also be a quotation (from an

English grammar-book perhaps).—"But suppose I *mean* it while I am saying it?" The old misunderstanding about the concept 'mean'.

394. "This is one of the things that I cannot doubt."

395. "I know all that." And that will come out in the way I act and in the way I speak about the things in question.

396. In the language-game (2),* can he say that he knows that those are building stones?—"No, but he *does* know it."

397. Haven't I gone wrong and isn't Moore perfectly right? Haven't I made the elementary mistake of confusing one's thoughts with one's knowledge? Of course I do not think to myself "The earth already existed for some time before my birth", but do I *know* it any the less? Don't I show that I know it by always drawing its consequences?

398. And don't I know that there is no stairway in this house going six floors deep into the earth, even though I have never thought about it?

399. But doesn't my drawing the consequences only show that I accept this hypothesis?

19.3.

400. Here I am inclined to fight windmills, because I cannot yet say the thing I really want to say.

401. I want to say: propositions of the form of empirical propo-

---

* *Philosophical Investigations* I§2. *Eds.*

sitions, and not only propositions of logic, form the foundation of all operating with thoughts (with language).— This observation is not of the form "I know . . .". "I know . . ." states what *I* know, and that is not of logical interest.

402. In this remark the expression "propositions of the form of empirical propositions" is itself thoroughly bad; the statements in question are statements about material objects. And they do not serve as foundations in the same way as hypotheses which, if they turn out to be false, are replaced by others.

. . . und schreib getrost
"Im Anfang war die Tat."*

403. To say of man, in Moore's sense, that he *knows* something; that what he says is therefore unconditionally the truth, seems wrong to me.—It is the truth only inasmuch as it is an unmoving foundation of his language-games.

404. I want to say: it's not that on some points men know the truth with perfect certainty. No: perfect certainty is only a matter of their attitude.

405. But of course there is still a mistake even here.

406. What I am aiming at is also found in the difference between the casual observation "I know that that's a . . .", as it might be used in ordinary life, and the same utterance when a philosopher makes it.

---

*   . . . and write with confidence
    "In the beginning was the deed."

                                        Goethe, *Faust* I. Trans.

407. For when Moore says "I know that that's . . ." I want to reply "you don't *know* anything!"—and yet I would not say that to anyone who was speaking without philosophical intention. That is, I feel (rightly?) that these two mean to say something different.

408. For if someone says he knows such-and-such, and this is part of his philosophy—then his philosophy is false if he has slipped up in this statement.

409. If I say "I know that that's a foot"—what am I really saying? Isn't the whole point that I am certain of the consequences—that if someone else had been in doubt I might say to him "you see—I told you so"? Would my knowledge still be worth anything if it let me down as a clue in action? And *can't* it let me down?

20.3.

410. Our knowledge forms an enormous system. And only within this system has a particular bit the value we give it.

411. If I say "*we assume* that the earth has existed for many years past" (or something similar), then of course it sounds strange that we should *assume* such a thing. But in the entire system of our language-games it belongs to the foundations. The assumption, one might say, forms the basis of action, and therefore, naturally, of thought.

412. Anyone who is unable to imagine a case in which one might say "I know that this is my hand" (and such cases are certainly rare) might say that these words were nonsense. True, he might also say "Of course I know—how

could I not know?"—but then he would possibly be taking the sentence "this is my hand" as an *explanation* of the words "my hand".

413. For suppose you were guiding a blind man's hand, and as you were guiding it along yours you said "this is my hand"; if he then said "are you sure?" or "do you know it is?", it would take very special circumstances for that to make sense.

414. But on the other hand: how do I *know* that it is my hand? Do I even here know exactly what it means to say it is my hand?—When I say "how do I know?" I do not mean that I have the least *doubt* of it. What we have here is a foundation for all my action. But it seems to me that it is wrongly expressed by the words "I know".

415. And in fact, isn't the use of the word "know" as a preeminently philosophical word altogether wrong? If "know" has this interest, why not "being certain"? Apparently because it would be too subjective. But isn't "know" *just* as subjective? Isn't one misled simply by the grammatical peculiarity that "p" follows from "I know p"?

   "I believe I know" would not need to express a lesser degree of certainty.—True, but one isn't trying to express even the greatest subjective certainty, but rather that certain propositions seem to underlie all questions and all thinking.

416. And have we an example of this in, say, the proposition that I have been living in this room for weeks past, that my memory does not deceive me in this?
   —"certain beyond all reasonable doubt"—

21.3.

417. "I know that for the last month I have had a bath every day." What am I remembering? Each day and the bath each morning? No. I *know* that I bathed each day and I do not derive that from some other immediate datum. Similarly I say "I felt a pain in my arm" without this locality coming into my consciousness in any other way (such as by means of an image).

418. Is my understanding only blindness to my own lack of understanding? It often seems so to me.

419. If I say "I have never been in Asia Minor", where do I get this knowledge from? I have not worked it out, no one told me; my memory tells me.—So I can't be wrong about it? Is there a truth here which I *know*?—I cannot depart from this judgment without toppling all other judgments with it.

420. Even a proposition like this one, that I am now living in England, has these two sides: it is not a *mistake*—but on the other hand, what do I know of England? Can't my judgment go all to pieces?

       Would it not be possible that people came into my room and all declared the opposite?—even gave me 'proofs' of it, so that I suddenly stood there like a madman alone among people who were all normal, or a normal person alone among madmen? Might I not then suffer doubts about what at present seems at the furthest remove from doubt?

421. I am in England.—Everything around me tells me so; wherever and however I let my thoughts turn, they con-

firm this for me at once.—But might I not be shaken if things such as I don't dream of at present were to happen?

422. So I am trying to say something that sounds like pragmatism.

Here I am being thwarted by a kind of *Weltanschauung*.

423. Then why don't I simply say with Moore "I *know* that I am in England"? Saying this is meaningful *in particular circumstances*, which I can imagine. But when I utter the sentence outside these circumstances, as an example to shew that I can know truths of this kind with certainty, then it at once strikes me as fishy.—Ought it to?

424. I say "I know p" either to assure people that I, too, know the truth p, or simply as an emphasis of ⊢p. One says, too, "I don't *believe* it, I *know* it". And one might also put it like this (for example): "That is a tree. And that's not just surmise."

But what about this: "If I were to tell someone that that was a tree, that wouldn't be just surmise." Isn't this what Moore was trying to say?

425. It would not be surmise and I might tell it to someone else with complete certainty, as something there is no doubt about. But does that mean that it is unconditionally the truth? May not the thing that I recognize with complete certainty as the tree that I have seen here my whole life long—may this not be disclosed as something different? May it not confound me?

And nevertheless it was right, in the circumstances that give this sentence meaning, to say "I know (I do not merely surmise) that that's a tree". To say that in strict truth I only believe it, would be wrong. It would be completely *misleading* to say: "I believe my name is L. W." And this too is right: I cannot be making a *mistake* about it. But that does not mean that I am infallible about it.

21.3.51

426. But how can we *show* someone that we *know* truths, not only about sense-data but also about things? For after all it can't be enough for someone to assure us that *he* knows this.

Well, what must our starting point be if we are to shew this?

22.3.

427. We need to shew that even if he never uses the words "I know . . .", his conduct exhibits the thing we are concerned with.

428. For suppose a person of normal behaviour assured us that he only *believed* his name was such-and-such, he *believed* he recognized the people he regularly lived with, he believed that he had hands and feet when he didn't actually see them, and so on. Can we shew him it is not so from the things he does (and says)?

23.3.51

429. What reason have I, now, when I cannot see my toes, to assume that I have five toes on each foot?

Is it right to say that my reason is that previous experience has always taught me so? Am I more certain of previous experience than that I have ten toes?

That previous experience may very well be the *cause* of my present certitude; but is it its ground?

430. I meet someone from Mars and he asks me "How many toes have human beings got?"—I say "Ten. I'll shew you", and take my shoes off. Suppose he was surprised that I knew with such certainty, although I hadn't looked at my toes—ought I to say: "We humans know how many toes we have whether we can see them or not"?

26.3.51

431. "I know that this room is on the second floor, that behind the door a short landing leads to the stairs, and so on." One could imagine cases where I should come out with this, but they would be extremely rare. But on the other hand I shew this knowledge day in, day out by my actions and also in what I say.

Now what does someone else gather from these actions and words of mine? Won't it be just that I am sure of my ground?—From the fact that I have been living here for many weeks and have gone up and down the stairs every day he will gather that I *know* where my room is situated.—I shall give him the assurance "I know" when he does *not* already know things which would have compelled the conclusion that I knew.

432. The utterance "I know . . ." can only have its meaning in connection with the other evidence of my 'knowing'.

433. So if I say to someone "I *know* that that's a tree", it is as if I
      told him "that is a tree; you can absolutely rely on it; there
      is no doubt about it". And a philosopher could only use the
      statement to show that this form of speech is actually used.
      But if his use of it is not to be merely an observation about
      English grammar, he must give the circumstances in which
      this expression functions.

434. Now does *experience* teach us that in such-and-such cir-
      cumstances people know this and that? Certainly, expe-
      rience shews us that normally after so-and-so many days
      a man can find his way about a house he has been liv-
      ing in. Or even: experience teaches us that after
      such-and-such a period of training a man's judgment is
      to be trusted. He must, experience tells us, have learnt
      for so long in order to be able to make a correct predic-
      tion. But ———

27.3.

435. One is often bewitched by a word. For example, by the
      word "know".

436. Is God bound by our knowledge? Are a lot of our state-
      ments *incapable* of falsehood? For that is what we want to
      say.

437. I am inclined to say: "That *cannot* be false." That is inter-
      esting; but what consequences has it?

438. It would not be enough to assure someone that I know
      what is going on at a certain place—without giving him
      grounds that satisfy him that I am in a position to know.

439. Even the statement "I know that behind this door there is a landing and the stairway down to the ground floor" only sounds so convincing because everyone takes it for granted that I know it.

440. There is something universal here; not just something personal.

441. In a court of law the mere assurance "I know . . ." on the part of a witness would convince no one. It must be shown that he was in a position to know.

Even the assurance "I know that that's a hand", said while someone looked at his own hand, would not be credible unless we knew the circumstances in which it was said. And if we do know them, it seems to be an assurance that the person speaking is normal in this respect.

442. For may it not happen that I *imagine* myself to *know* something?

443. Suppose that in a certain language there were no word corresponding to our "know".—The people simply make assertions. ("That is a tree", etc.) Naturally it can occur for them to make mistakes. And so they attach a sign to the sentence which indicates how probable they take a mistake to be—or should I say, how probable a mistake is in this case? This latter can also be indicated by mentioning certain circumstances. For example "Then A said to B '. . .'. I was standing quite close to them and my hearing is good", or "A was at such-and-such a place yesterday. I saw him from a long way off. My eyes are not very good", or "There is a tree over there: I can see it clearly and I have seen it innumerable times before".

444. "The train leaves at two o'clock. Check it once more to make certain" or "The train leaves at two o'clock. I have just looked it up in a new time-table". One may also add "I am reliable in such matters". The usefulness of such additions is obvious.

445. But if I say "I have two hands", what can I add to indicate reliability? At the most that the circumstances are the ordinary ones.

446. But why *am* I so certain that this is my hand? Doesn't the whole language-game rest on this kind of certainty?

   Or: isn't this 'certainty' already presupposed in the language-game? Namely by virtue of the fact that one is not playing the game, or is playing it wrong, if one does not recognize objects with certainty.

28.3.

447. Compare with this $12 \times 12 = 144$. Here too we don't say "perhaps". For, in so far as this proposition rests on our not miscounting or miscalculating and on our senses not deceiving us as we calculate, both propositions, the arithmetical one and the physical one, are on the same level.

   I want to say: The physical game is just as certain as the arithmetical. But this can be misunderstood. My remark is a logical and not a psychological one.

448. I want to say: If one doesn't marvel at the fact that the propositions of arithmetic (e.g. the multiplication tables) are 'absolutely certain', then why should one be astonished that the proposition "This is my hand" is so equally?

449. Something must be taught us as a foundation.

450. I want to say: our learning has the form "that is a violet", "that is a table". Admittedly, the child might hear the word "violet" for the first time in the sentence "perhaps that is a violet", but then he could ask "what is a violet?" Now this might of course be answered by showing him a picture. But how would it be if one said "that is a . . ." only when showing him a picture, but otherwise said nothing but "perhaps that is a . . ."—What practical consequences is that supposed to have?

A doubt that doubted everything would not be a doubt.

451. My objection against Moore, that the meaning of the isolated sentence "That is a tree" is undetermined, since it is not determined what the "*that*" is that is said to be a tree—doesn't work, for one can make the meaning more definite by saying, for example: "The object over there that looks like a tree is not an artificial imitation of a tree but a real one."

452. It would not be reasonable to doubt if that was a real tree or only. . . .

My finding it beyond doubt is not what counts. If a doubt would be unreasonable, that cannot be seen from what *I* hold. There would therefore have to be a rule that declares doubt to be unreasonable here. But there isn't such a rule, either.

453. I do indeed say: "Here no reasonable person would doubt."—Could we imagine learned judges being asked whether a doubt was reasonable or unreasonable?

454. There are cases where doubt is unreasonable, but others where it seems logically impossible. And there seems to be no clear boundary between them.

29.3.

455. Every language-game is based on words 'and objects' being recognized again. We learn with the same inexorability that this is a chair as that $2 \times 2 = 4$.

456. If, therefore, I doubt or am uncertain about this being my hand (in whatever sense), why not in that case about the meaning of these words as well?

457. Do I want to say, then, that certainty resides in the nature of the language-game?

458. One doubts on specific grounds. The question is this: how is doubt introduced into the language-game?

459. If the shopkeeper wanted to investigate each of his apples without any reason, for the sake of being certain about everything, why doesn't he have to investigate the investigation? And can one talk of belief here (I mean belief as in 'religious belief', not surmise)? All psychological terms merely distract us from the thing that really matters.

460. I go to the doctor, shew him my hand and say "This is a hand, not . . .; I've injured it, etc., etc." Am I only giving him a piece of superfluous information? For example, mightn't one say: supposing the words "This is a hand" *were* a piece of information—how could you bank on his understanding this information? Indeed, if it is open

to doubt 'whether that is a hand', why isn't it also open to doubt whether I am a human being who is informing the doctor of this?—But on the other hand one can imagine cases—even if they are very rare ones—where this declaration is not superfluous; or is only superfluous but not absurd.

461. Suppose that I were the doctor and a patient came to me, showed me his hand and said: "This thing that looks like a hand isn't just a superb imitation—it really is a hand" and went on to talk about his injury—should I really take this as a piece of information, even though a superfluous one? Shouldn't I be more likely to consider it nonsense, which admittedly did have the form of a piece of information? For, I should say, if this information really were meaningful, how can he be certain of what he says? The background is lacking for it to be information.

30.3.

462. Why doesn't Moore produce as one of the things that he knows, for example, that in such-and-such a part of England there is a village called so-and-so? In other words: why doesn't he mention a fact that is known to him and not to *every one* of us?

31.3.

463. This is certainly true, that the information "That is a tree", when no one could doubt it, might be a kind of joke and as such have meaning. A joke of this kind was in fact made once by Renan.

3.4.51

464. My difficulty can also be shewn like this: I am sitting talking to a friend. Suddenly I say: "I knew all along that you
were so-and-so." Is that really just a superfluous, though
true, remark?

  I feel as if these words were like "Good morning" said
to someone in the middle of a conversation.

465. How would it be if we had the words "They know nowa-
days that there are over . . . species of insects" instead of "I
know that that's a tree"? If someone were suddenly to utter
the first sentence out of all context one might think: he has
been thinking of something else in the interim and is now
saying out loud some sentence in his train of thought. Or
again: he is in a trance and is speaking without understanding what he is saying.

466. Thus it seems to me that I have known something the
whole time, and yet there is no meaning in saying so, in
uttering this truth.

467. I am sitting with a philosopher in the garden; he says again
and again "I know that that's a tree", pointing to a tree
that is near us. Someone else arrives and hears this, and I
tell him: "This fellow isn't insane. We are only doing philosophy."

4.4.

468. Someone says irrelevantly "That's a tree". He might say
this sentence because he remembers having heard it in a
similar situation; or he was suddenly struck by the tree's
beauty and the sentence was an exclamation; or he was

pronouncing the sentence to himself as a grammatical example; etc., etc. And now I ask him "How did you mean that?" and he replies "It was a piece of information directed at you". Shouldn't I be at liberty to assume that he doesn't know what he is saying, if he is insane enough to want to give me this information?

469. In the middle of a conversation, someone says to me out of the blue: "I wish you luck." I am astonished; but later I realize that these words connect up with his thoughts about me. And now they do not strike me as meaningless any more.

470. Why is there no doubt that I am called L. W.? It does not seem at all like something that one could establish at once beyond doubt. One would not think that it is one of the indubitable truths.

5.4.

[Here there is still a big gap in my thinking. And I doubt whether it will be filled now.]

471. It is so difficult to find the *beginning*. Or, better: it is difficult to begin at the beginning. And not try to go further back.

472. When a child learns language it learns at the same time what is to be investigated and what not. When it learns that there is a cupboard in the room, it isn't taught to doubt whether what it sees later on is still a cupboard or only a kind of stage set.

473. Just as in writing we learn a particular basic form of letters and then vary it later, so we learn first the stability of things as the norm, which is then subject to alterations.

474. This game proves its worth. That may be the cause of its being played, but it is not the ground.

475. I want to regard man here as an animal; as a primitive being to which one grants instinct but not ratiocination. As a creature in a primitive state. Any logic good enough for a primitive means of communication needs no apology from us. Language did not emerge from some kind of ratiocination.

6.4.

476. Children do not learn that books exist, that armchairs, exist, etc. etc.,—they learn to fetch books, sit in armchairs, etc. etc.

Later, questions about the existence of things do of course arise. "Is there such a thing as a unicorn?" and so on. But such a question is possible only because as a rule no corresponding question presents itself. For how does one know how to set about satisfying oneself of the existence of unicorns? How did one learn the method for determining whether something exists or not?

477. "So one must know that the objects whose names one teaches a child by an ostensive definition exist."—Why must one know they do? Isn't it enough that experience doesn't later show the opposite?

For why should the language-game rest on some kind of knowledge?

7.4.

478. Does a child believe that milk exists? Or does it know that milk exists? Does a cat know that a mouse exists?

479. Are we to say that the knowledge that there are physical objects comes very early or very late?

8.4.

480. A child that is learning to use the word "tree". One stands with it in front of a tree and says "*Lovely* tree!" Clearly no doubt as to the tree's existence comes into the language-game. But can the child be said to *know*: 'that a tree exists'? Admittedly it's true that 'knowing something' doesn't involve *thinking* about it—but mustn't anyone who knows something be capable of doubt? And doubting means thinking.

481. When one hears Moore say "I *know* that that's a tree", one suddenly understands those who think that that has by no means been settled.

The matter strikes one all at once as being unclear and blurred. It is as if Moore had put it in the wrong light.

It is as if I were to see a painting (say a painted stage-set) and recognize what it represents from a long way off at once and without the slightest doubt. But now I step nearer: and then I see a lot of patches of different colours, which are all highly ambiguous and do not provide any certainty whatever.

482. It is as if "I know" did not tolerate a metaphysical emphasis.

483. The correct use of the expression "I know". Someone with bad sight asks me: "do you believe that the thing we can see there is a tree?" I reply "I *know* it is; I can see it clearly and am familiar with it".—A: "Is N. N. at home?"—I: "I believe he is."—A: "Was he at home yesterday?"—I: "Yesterday he was—I know he was; I spoke to him."—A: "Do

you know or only believe that this part of the house is built on later than the rest?"—I: "I *know* it is; I asked so and so about it."

484. In these cases, then, one says "I know" and mentions how one knows, or at least one can do so.

485. We can also imagine a case where someone goes through a list of propositions and as he does so keeps asking "Do I know that or do I only believe it?" He wants to check the certainty of each individual proposition. It might be a question of making a statement as a witness before a court.

9.4.

486. "Do you know or do you only believe that your name is L. W.?" Is that a meaningful question?

Do you know or do you only believe that what you are writing down now are German words? Do you only believe that "believe" has *this* meaning? *What* meaning?

487. What is the proof that I *know* something? Most certainly not my saying I know it.

488. And so, when writers enumerate all the things they *know*, that proves nothing whatever.

So the possibility of knowledge about physical objects cannot be proved by the protestations of those who believe that they have such knowledge.

489. For what reply does one make to someone who says "I believe it merely strikes you as if you knew it"?

490. When I ask "Do I know or do I only believe that I am called . . . ?" it is no use to look within myself.

But I could say: not only do I never have the slightest doubt that I am called that, but there is no judgment I could be certain of if I started doubting about that.

10.4.

491. "Do I know or do I only believe that I am called L. W.?"—Of course, if the question were "Am I certain or do I only surmise . . . ?", then my answer could be relied on.

492. "Do I know or do I only believe . . . ?" might also be expressed like this: What if it *seemed* to turn out that what until now has seemed immune to doubt was a false assumption? Would I react as I do when a belief has proved to be false? or would it seem to knock from under my feet the ground on which I stand in making any judgments at all?—But of course I do not intend this as a *prophecy*.

Would I simply say "I should never have thought it!"— or would I (have to) refuse to revise my judgment—because such a 'revision' would amount to annihilation of all yardsticks?

493. So is this it: I must recognize certain authorities in order to make judgments at all?

494. "I cannot doubt this proposition without giving up all judgment."

But what sort of proposition is that? (It is reminiscent of what Frege said about the law of identity.*) It is cer-

---

* *Grundgesetze der Arithmetik* I xviii *Eds.*

tainly no empirical proposition. It does not belong to psychology. It has rather the character of a rule.

495. One might simply say "O, rubbish!" to someone who wanted to make objections to the propositions that are beyond doubt. That is, not reply to him but admonish him.

496. This is a similar case to that of shewing that it has no meaning to say that a game has always been played wrong.

497. If someone wanted to arouse doubts in me and spoke like this: here your memory is deceiving you, there you've been taken in, there again you have not been thorough enough in satisfying yourself, etc., and if I did not allow myself to be shaken but kept to my certainty—then my doing so cannot be wrong, even if only because this is just what defines a game.

11.4.

498. The queer thing is that even though I find it quite correct for someone to say "Rubbish!" and so brush aside the attempt to confuse him with doubts at bedrock,—nevertheless, I hold it to be incorrect if he seeks to defend himself (using, e.g., the words "I know").

499. I might also put it like this: the 'law of induction' can no more be *grounded* than certain particular propositions concerning the material of experience.

500. But it would also strike me as nonsense to say "I know that the law of induction is true".

Imagine such a statement made in a court of law! It

would be more correct to say "I believe in the law of . . ." where 'believe' has nothing to do with *surmising*.

501. Am I not getting closer and closer to saying that in the end logic cannot be described? You must look at the practice of language, then you will see it.

502. Could one say "I know the position of my hands with my eyes closed", if the position I gave always or mostly contradicted the evidence of other people?

503. I look at an object and say "That is a tree", or "I know that that's a tree".—Now if I go nearer and it turns out that it isn't, I may say "It wasn't a tree after all" or alternatively I say "It *was* a tree but now it isn't any longer". But if all the others contradicted me, and said it never had been a tree, and if all the other evidences spoke against me—what *good* would it do me to stick to my "I know"?

504. Whether I *know* something depends on whether the evidence backs me up or contradicts me. For to say one knows one has a pain means nothing.

505. It is always by favour of Nature that one knows something.

506. "If my memory deceives me *here* it can deceive me everywhere."
    If I don't know *that*, how do I know if my words mean what I believe they mean?

507. "If this deceives me, what does 'deceive' mean any more?"

508. What can I rely on?

509. I really want to say that a language-game is only possible if one trusts something (I did not say "can trust something").

510. If I say "Of course I know that that's a towel" I am making an *utterance*.* I have no thought of a verification. For me it is an immediate utterance.

   I don't think of past or future. (And of course it's the same for Moore, too.)

   It is just like directly taking hold of something, as I take hold of my towel without having doubts.

511. And yet this direct taking-hold corresponds to a *sureness*, not to a knowing.

   But don't I take hold of a thing's name like that, too?

12.4.

512. Isn't the question this: "What if you had to change your opinion even on these most fundamental things?" And to that the answer seems to me to be: "You don't *have* to change it. That is just what their being 'fundamental' is."

513. What if something *really unheard-of* happened?—If I, say, saw houses gradually turning into steam without any obvious cause, if the cattle in the fields stood on their heads and laughed and spoke comprehensible words; if trees gradually changed into men and men into trees. Now, was I right when I said before all these things happened "I know that that's a house" etc., or simply "that's a house" etc.?

---

* *Äußerung* (Eds.)

514. This statement appeared to me fundamental; if it is false, what are 'true' or 'false' any more?!

515. If my name is *not* L. W., how can I rely on what is meant by "true" and "false"?

516. If something happened (such as someone telling me something) calculated to make me doubtful of my own name, there would certainly also be something that made the grounds of these doubts themselves seem doubtful, and I could therefore decide to retain my old belief.

517. But might it not be possible for something to happen that threw me entirely off the rails? Evidence that made the most certain thing unacceptable to me? Or at any rate made me throw over my most fundamental judgments? (Whether rightly or wrongly is beside the point.)

518. Could I imagine observing this in another person?

519. Admittedly, if you are obeying the order "Bring me a book", you may have to check whether the thing you see over there really is a book, but then you do at least know what people mean by "book"; and if you don't you can look it up,—but then you must know what some other word means. And the fact that a word means such-and-such, is used in such-and-such a way, is in turn an empirical fact, like the fact that what you see over there is a book.

Therefore, in order for you to be able to carry out an order there must be some empirical fact about which you are not in doubt. Doubt itself rests only on what is beyond doubt.

But since a language-game is something that consists

in the recurrent procedures of the game in time, it seems impossible to say in any *individual* case that such-and-such must be beyond doubt if there is to be a language-game— though it is right enough to say that *as a rule* some empirical judgment or other must be beyond doubt.

13.4.

520. Moore has every right to say he knows there's a tree there in front of him. Naturally he may be wrong. (For it is *not* the same as with the utterance "I believe there is a tree there".) But whether he is right or wrong in this case is of no philosophical importance. If Moore is attacking those who say that one cannot really know such a thing, he can't do it by assuring them that *he* knows this and that. For one need not believe him. If his opponents had asserted that one could not *believe* this and that, then he could have replied: "*I* believe it."

14.4.

521. Moore's mistake lies in this—countering the assertion that one cannot know that, by saying "I do know it".

522. We say: if a child has mastered language—and hence its application—it must know the meaning of words. It must, for example, be able to attach the name of its colour to a white, black, red or blue object without the occurrence of any doubt.

523. And indeed no one misses doubt here; no one is surprised that we do not merely *surmise* the meaning of our words.

15.4.

524. Is it essential for our language-games ('ordering and obeying' for example) that no doubt appears at certain points,
or is it enough if there is the feeling of being sure, admittedly with a slight breath of doubt?

That is, is it enough if I do not, as I do now, call something 'black', 'green', 'red', *straight off*, without any doubt
at all interposing itself—but do instead say "I am sure that
that is red", as one may say "I am sure that he will come
today" (in other words with the 'feeling of being sure')?

The accompanying feeling is of course a matter of indifference to us, and equally we have no need to bother
about the words "I am sure that" either.—What is important is whether they go with a difference in the *practice* of
the language.

One might ask whether a person who spoke like this
would always say "I am sure" on occasions where (for example) there is sureness in the reports we make (in an experiment, for example, we look through a tube and report
the colour we see through it). If he does, our immediate
inclination will be to check what he says. But if he proves
to be perfectly reliable, one will say that his way of talking
is merely a bit perverse, and does not affect the issue. One
might for example suppose that he has read sceptical philosophers, become convinced that one can know nothing,
and that is why he has adopted this way of speaking. Once
we are used to it, it does not infect practice.

525. What, then, does the case look like where someone really
has got a different relationship to the names of colours, for
example, from us? Where, that is, there persists a slight
doubt or a possibility of doubt in their use.

16.4.

526. If someone were to look at an English pillar-box and say "I
     am sure that it's red", we should have to suppose that he
     was colour-blind, or believe he had no mastery of English
     and knew the correct name for the colour in some other
     language.

     If neither was the case we should not quite understand
     him.

527. An Englishman who calls this colour "red" is not: 'sure it
     is called "red" in English'.

     A child who has mastered the use of the word is not
     'sure that in his language this colour is called . . .'. Nor can
     one say of him that when he is learning to speak he learns
     that the colour is called that in English; nor yet: he *knows*
     this when he has learnt the use of the word.

528. And in spite of this: if someone asked me what the colour
     was called in German and I tell him, and now he asks me
     "are you sure?"—then I shall reply "I *know* it is; German is
     my mother tongue".

529. And one child, for example, will say, of another or of
     himself, that he already knows what such-and-such is
     called.

530. I may tell someone "this colour is called 'red' in English"
     (when for example I am teaching him English). In this
     case I should not say "I know that this colour . . ."—I
     would perhaps say that if I had just now learned it, or by
     contrast with another colour whose English name I am
     not acquainted with.

531. But now, isn't it correct to describe my present state as follows: I *know* what this colour is called in English? And if that is correct, why then should I not describe my state with the corresponding words "I know etc."?

532. So when Moore sat in front of a tree and said "I know that that's a tree", he was simply stating the truth about his state at the time.

[I do philosophy now like an old woman who is always mislaying something and having to look for it again: now her spectacles, now her keys.]

533. Well, if it was correct to describe his state out of context, then it was just as correct to utter the words "that's a tree" out of context.

534. But is it wrong to say: "A child that has mastered a language-game must *know* certain things"?

If instead of that one said "must be *able to do* certain things", that would be a pleonasm, yet this is just what I want to counter the first sentence with.—But: "a child acquires a knowledge of natural history". That presupposes that it can ask what such and such a plant is called.

535. The child knows what something is called if he can reply correctly to the question "what is that called?"

536. Naturally, the child who is just learning to speak has not yet got the concept *is called* at all.

537. Can one say of someone who hasn't this concept that he *knows* what such-and-such is called?

538. The child, I should like to say, learns to react in such-and-such a way; and in so reacting it doesn't so far know anything. Knowing only begins at a later level.

539. Does it go for knowing as it does for collecting?

540. A dog might learn to run to N at the call "N", and to M at the call "M",—but would that mean he knows what these people are called?

541. "He only knows what this person is called—not yet what that person is called". That is something one cannot, strictly speaking, say of someone who simply has not yet got the concept of people's having names.

542. "I can't describe this flower if I don't know that this colour is called 'red'."

543. A child can use the names of people long before he can say in any form whatever: "I know this one's name; I don't know that one's yet."

544. Of course I may truthfully say "I know what this colour is called in English", at the same time as I point (for example) to the colour of fresh blood. But — — —

17.4.

545. 'A child knows which colour is meant by the word "blue".' What he knows here is not all that simple.

546. I should say "I know what this colour is called" if e.g. what is in question is shades of colour whose name not everybody knows.

547. One can't yet say to a child who is just beginning to speak and can use the words "red" and "blue": "Come on, you know what this colour is called!"

548. A child must learn the use of colour words before it can ask for the name of a colour.

549. It would be wrong to say that I can only say "I know that there is a chair there" when there is a chair there. Of course it isn't *true* unless there is, but I have a right to say this if I am *sure* there is a chair there, even if I am wrong.

   [Pretensions are a mortgage which burdens a philosopher's capacity to think.]

18.4.

550. If someone believes something, we needn't always be able to answer the question 'why he believes it'; but if he knows something, then the question "how does he know?" must be capable of being answered.

551. And if one does answer this question, one must do so according to generally accepted axioms. *This* is how something of this sort may be known.

552. Do I know that I am now sitting in a chair?—Don't I know it?! In the present circumstances no one is going to say that I know this; but no more will he say, for example, that I am conscious. Nor will one normally say this of the passers-by in the street.

   But now, even if one doesn't say it, does that make it *untrue*??

553. It is queer: if I say, without any special occasion, "I know"—for example, "I know that I am now sitting in a chair", this statement seems to me unjustified and presumptuous. But if I make the same statement where there is some need for it, then, although I am not a jot more certain of its truth, it seems to me to be perfectly justified and everyday.

554. In its language-game it is not presumptuous. There, it has no higher position than, simply, the human language-game. For there it has its restricted application.

But as soon as I say this sentence outside its context, it appears in a false light. For then it is as if I wanted to insist that there are things that I *know*. God himself can't say anything to me about them.

19.4

555. We say we know that water boils when it is put over a fire. How do we know? Experience has taught us.—I say "I know that I had breakfast this morning"; experience hasn't taught me that. One also says "I know that he is in pain". The language-game is different every time, we are *sure* every time, and people will agree with us that we are *in a position* to know every time. And that is why the propositions of physics are found in textbooks for everyone.

If someone says he *knows* something, it must be something that, by general consent, he is in a position to know.

556. One doesn't say: he is in a position to believe that.

But one does say: "It is reasonable to assume that in this situation" (or "to believe that").

557. A court-martial may well have to decide whether it was reasonable in such-and-such a situation to have assumed this or that with confidence (even though wrongly).

558. We say we know that water boils and does not freeze under such-and-such circumstances. Is it conceivable that we are wrong? Wouldn't a mistake topple all judgment with it? More: what could stand if that were to fall? Might someone discover something that made us say "It was a mistake"?

    Whatever may happen in the future, however water may behave in the future,—we *know* that up to now it has behaved *thus* in innumerable instances.

    This fact is fused into the foundations of our language-game.

559. You must bear in mind that the language-game is so to say something unpredictable. I mean: it is not based on grounds. It is not reasonable (or unreasonable).

    It is there—like our life.

560. And the concept of knowing is coupled with that of the language-game.

561. "I know" and "You can rely on it". But one cannot always substitute the latter for the former.

562. At any rate it is important to imagine a language in which *our* concept 'knowledge' does not exist.

563. One says "I know that he is in pain" although one can produce no convincing grounds for this.—Is this the same as "I am sure that he . . ."?—No. "I am sure" tells you my

subjective certainty. "I know" means that I who know it, and the person who doesn't are separated by a difference in understanding. (Perhaps based on a difference in degree of experience.)

If I say "I know" in mathematics, then the justification for this is a proof.

If in these two cases instead of "I know", one says "you can rely on it" then the substantiation is of a different kind in each case.

And substantiation comes to an end.

564. A language-game: bringing building stones, reporting the number of available stones. The number is sometimes estimated, sometimes established by counting. Then the question arises "Do you believe there are as many stones as that?", and the answer "I know there are—I've just counted them". But here the "I know" could be dropped. If, however, there are several ways of finding something out for sure, like counting, weighing, measuring the stack, then the statement "I know" can take the place of mentioning *how* I know.

565. But here there isn't yet any question of any 'knowledge' that *this* is called "a slab", *this* "a pillar", etc.

566. Nor does a child who learns my language-game (No. 2)* learn to say "I know that this is called 'a slab'".

Now of course there is a language-game in which the child uses *that* sentence. This presupposes that the child is already capable of using the name as soon as he is given it. (As if someone were to tell me "this colour is called . . .".)—

---

* *Philosophical Investigations* §2. Eds.

Thus, if the child has learnt a language-game with building stones, one can say something like "and *this* stone is called '. . .'", and in this way the original language-game has been *expanded*.

567. And now, is my knowledge that I am called L. W. of the same kind as knowledge that water boils at 100°C.? Of course, this question is wrongly put.

568. If one of my names were used only very rarely, then it might happen that I did not know it. It goes without saying that I know my name, only because, like anyone else, I use it over and over again.

569. An inner experience cannot shew me that I *know* something.
Hence, if in spite of that I say, "I know that my name is . . .", and yet it is obviously not an empirical proposition,

———

570. "I know this is my name; among us any grown-up knows what his name is."

571. "My name is . . .—you can rely on that. If it turns out to be wrong you need never believe me in the future."

572. Don't I seem to know that I can't be wrong about such a thing as my own name?
This comes out in the words: "If that is wrong, then I am crazy." Very well, but those are words; but what influence has it on the application of language?

573. Is it through the impossibility of anything's convincing me of the contrary?

574. The question is, what *kind* of proposition is: "I know I can't be mistaken about that", or again "I can't be mistaken about that"?

      This "I know" seems to prescind from all grounds: I simply *know* it. But if there can be any question at all of being mistaken here, then it must be possible to test whether I know it.

575. Thus the purpose of the phrase "I know" might be to indicate where I can be relied on; but where that's what it's doing, the usefulness of this sign must emerge from *experience*.

576. One might say "How do I know that I'm not mistaken about my name?"—and if the reply was "Because I have used it so often", one might go on to ask "How do I know that I am not mistaken about *that*?" And here the "How do I know" cannot any longer have any significance.

577. "My knowledge of my name is absolutely definite."

      I would refuse to entertain any argument that tried to show the opposite!

      And what does "I *would* refuse" mean? Is it the expression of an intention?

578. But mightn't a higher authority assure me that I don't know the truth? So that I had to say "Teach me!"? But then my eyes would have to be opened.

579. It is part of the language-game with people's names that everyone knows his name with the greatest certainty.

20.4.

580. It might surely happen that whenever I said "I know" it turned out to be wrong. (Shewing up.)

581. But perhaps I might nevertheless be unable to help myself, so that I kept on declaring "I know . . .". But ask yourself: how did the child learn the expression?

582. "I know that" may mean: I am quite familiar with it—or again: it is certainly so.

583. "I know that the name of this in . . . is '. . . '"—How do you know?—"I have learnt . . .".

     Could I substitute "In . . . the name of this is '. . . '" for "I know etc." in this example?

584. Would it be possible to make use of the verb "know" only in the question "How do you know?" following a simple assertion?—Instead of "I already know that" one says "I am familiar with that"; and this follows only upon being told the fact. But* what does one say instead of "I know what that is"?

585. But doesn't "I know that that's a tree" say something different from "that is a tree"?

586. Instead of "I know what that is" one might say "I can say what that is". And if one adopted this form of expression what would then become of "I know that that is a . . ."?

---

* The last sentence is a later addition. Eds.

587. Back to the question whether "I know that that's a . . ." says anything different from "that is a . . .". In the first sentence a person is mentioned, in the second, not. But that does not shew that they have different meanings. At all events one often replaces the first form by the second, and then often gives the latter a special intonation. For one speaks differently when one makes an uncontradicted assertion from when one maintains an assertion in face of contradiction.

588. But don't I use the words "I know that . . ." to say that I am in a certain state, whereas the mere assertion "that is a . . ." does not say this? And yet one often does reply to such an assertion by asking "how do you know?"—"But surely, only because the fact that I assert this gives to understand that I think I know it".—This point could be made in the following way: In a zoo there might be a notice "this is a zebra"; but never "I know that this is a zebra".

   "I know" has meaning only when it is uttered by a person. But, given that, it is a matter of indifference whether what is uttered is "I know . . ." or "That is . . .".

589. For how does a man learn to recognize his own state of knowing something?

590. At most one might speak of recognizing a state, where what is said is "I know what that is". Here one can satisfy oneself that one really is in possession of this knowledge.

591. "I know what kind of tree that is.—It is a chestnut."
   "I know what kind of tree that is.—I know it's a chestnut."
   The first statement sounds more natural than the second. One will only say "I know" a second time if one

wants especially to emphasize certainty; perhaps to anticipate being contradicted. The first "I know" means roughly:
I can say.

But in another case one might begin with the observation "that's a . . .", and then, when this is contradicted,
counter by saying: "I know what sort of a tree it is", and by
this means lay emphasis on being sure.

592. "I can tell you what kind of a . . . that is, and no doubt
about it."

593. Even when one can replace "I know" by "It is . . ." still one
cannot replace the negation of the one by the negation of
the other.

With "I don't know . . ." a new element enters our
language-games.

21.4.

594. My name is "L. W." And if someone were to dispute it, I
should straightaway make connexions with innumerable
things which make it certain.

595. "But I can still imagine someone making all these connexions, and none of them corresponding with reality. Why
shouldn't I be in a similar case?"

If I imagine such a person I also imagine a reality, a
world that surrounds him; and I imagine him as thinking
(and speaking) in contradiction to this world.

596. If someone tells me his name is N. N., it is meaningful for
me to ask him "Can you be mistaken?" That is an allow-
able question in the language-game. And the answer to it,
yes or no, makes sense.—Now of course this answer is not

infallible either, i.e., there might be a time when it proved to be wrong, but that does not deprive the question "Can you be . . ." and the answer "No" of their meaning.

597. The reply to the question "Can you be mistaken?" gives the statement a definite weight. The answer may also be: "I don't *think* so."

598. But couldn't one reply to the question "Can you . . ." by saying: "I will describe the case to you and then you can judge for yourself whether I can be mistaken"?

For example, if it were a question of someone's own name, the fact might be that he had never used this name, but remembered he had read it on some document,—but on the other hand the answer might be: "I've had this name my whole life long, I've been called it by everybody." If *that* is not equivalent to the answer "I can't be mistaken", then the latter has no meaning whatever. And yet quite obviously it points to a very important distinction.

599. For example one could describe the certainty of the proposition that water boils at *circa* 100°C. That isn't e.g. a proposition I have once heard (like this or that, which I could mention). I made the experiment myself at school. The proposition is a very elementary one in our text-books, which are to be trusted in matters like this because . . . —Now one can offer counter-examples to all this, which show that human beings have held this and that to be certain which later, according to our opinion, proved false. But the argument is worthless.* To say: in the end we can

---

* *Marginal note.* May it not also happen that we believe we recognize a mistake of earlier times and later come to the conclusion that the first opinion was the right one? etc.

only adduce such grounds as *we* hold to be grounds, is to say nothing at all.

I believe that at the bottom of this is a misunderstanding of the nature of our language-games.

600. What kind of grounds have I for trusting text-books of experimental physics?

I have no grounds for not trusting them. And I trust them. I know how such books are produced—or rather, I believe I know. I have some evidence, but it does not go very far and is of a very scattered kind. I have heard, seen and read various things.

22.4.

601. There is always the danger of wanting to find an expression's meaning by contemplating the expression itself, and the frame of mind in which one uses it, instead of always thinking of the practice. That is why one repeats the expression to oneself so often, because it is as if one must see what one is looking for in the expression and in the feeling it gives one.

23.4.

602. Should I say "I believe in physics", or "I know that physics is true"?

603. I am taught that under *such* circumstances *this* happens. It has been discovered by making the experiment a few times. Not that that would prove anything to us, if it weren't that this experience was surrounded by others which combine with it to form a system. Thus, people did not make experiments just about falling bodies but also about air resistance and all sorts of other things.

But in the end I rely on these experiences, or on the reports of them, I feel no scruples about ordering my own activities in accordance with them.—But hasn't this trust also proved itself? So far as I can judge—yes.

604. In a court of law the statement of a physicist that water boils at about 100°C. would be accepted unconditionally as truth.

If I mistrusted this statement what could I do to undermine it? Set up experiments myself? What would they prove?

605. But what if the physicist's statement were superstition and it were just as absurd to go by it in reaching a verdict as to rely on ordeal by fire?

606. That to my mind someone else has been wrong is no ground for assuming that I am wrong now.—But isn't it a ground for assuming that I *might* be wrong? It is *no* ground for any *unsureness* in my judgment, or my actions.

607. A judge might even say "That is the truth—so far as a human being can know it". But what would this rider achieve? ("beyond all reasonable doubt").

608. Is it wrong for me to be guided in my actions by the propositions of physics? Am I to say I have no good ground for doing so? Isn't precisely this what we call a 'good ground'?

609. Supposing we met people who did not regard that as a telling reason. Now, how do we imagine this? Instead of the physicist, they consult an oracle. (And for that we consider them primitive.) Is it wrong for them to consult an oracle

and be guided by it?—If we call this "wrong" aren't we using our language-game as a base from which to *combat* theirs?

610. And are we right or wrong to combat it? Of course there are all sorts of slogans which will be used to support our proceedings.

611. Where two principles really do meet which cannot be reconciled with one another, then each man declares the other a fool and heretic.

612. I said I would 'combat' the other man,—but wouldn't I give him *reasons*? Certainly; but how far do they go? At the end of reasons comes *persuasion*. (Think what happens when missionaries convert natives.)

613. If I now say "I know that the water in the kettle on the gas-flame will not freeze but boil", I seem to be as justified in this "I know" as I am in *any*. 'If I know anything I know *this*'.—Or do I know with still *greater* certainty that the person opposite me is my old friend so-and-so? And how does that compare with the proposition that I am seeing with two *eyes* and shall see them if I look in the glass?—I don't know confidently what I am to answer here.—But still there is a difference between the cases. If the water over the gas freezes, of course I shall be as astonished as can be, but I shall assume some factor I don't know of, and perhaps leave the matter to physicists to judge. But what could make me doubt whether this person here is N. N., whom I have known for years? Here a doubt would seem to drag everything with it and plunge it into chaos.

614. That is to say: If I were contradicted on all sides and told that this person's name was not what I had always known it was (and I use "know" here intentionally), then in that case the foundation of all judging would be taken away from me.

615. Now does that mean: "I can only make judgments at all because things behave thus and thus (as it were, behave kindly)"?

616. Why, would it be *unthinkable* that I should stay in the saddle however much the facts bucked?

617. Certain events would put me into a position in which I could not go on with the old language-game any further. In which I was torn away from the *sureness* of the game.

   Indeed, doesn't it seem obvious that the possibility of a language-game is conditioned by certain facts?

618. In that case it would seem as if the language-game must '*show*' the facts that make it possible. (But that's not how it is.)

   Then can one say that only a certain regularity in occurrences makes induction possible? The 'possible' would of course have to be '*logically possible*'.

619. Am I to say: even if an irregularity in natural events did suddenly occur, that wouldn't *have* to throw me out of the saddle. I might make inferences then just as before, but whether one would call that "induction" is another question.

620. In particular circumstances one says "you can rely on this"; and this assurance may be justified or unjustified in everyday language, and it may also count as justified even when

what was foretold does not occur. *A language-game exists* in which this assurance is employed.

24.4.

621. If anatomy were under discussion I should say: "I know that twelve pairs of nerves lead from the brain." I have never seen these nerves, and even a specialist will only have observed them in a few specimens.—This just is how the word "know" is correctly used here.

622. But now it is also correct to use "I know" in the contexts which Moore mentioned, at least *in particular circumstances*. (Indeed, I do not know what "I know that I am a human being" means. But even that might be given a sense.)

For each one of these sentences I can imagine circumstances that turn it into a move in one of our language-games, and by that it loses everything that is philosophically astonishing.

623. What is odd is that in such a case I always feel like saying (although it is wrong): "I know that—so far as one can know such a thing." That is incorrect, but something right is hidden behind it.

624. "Can you be mistaken about this colour's being called 'green' in English?" My answer to this can only be "No". If I were to say "Yes, for there is always the possibility of a delusion", that would mean nothing at all.

For is that rider something unknown to the other? And how is it known to me?

625. But does that mean that it is unthinkable that the word "green" should have been produced here by a slip of the

tongue or a momentary confusion? Don't we know of such cases?—One can also say to someone "Mightn't you perhaps have made a slip?" That amounts to: "Think about it again".—

But these rules of caution only make sense if they come to an end somewhere.

A doubt without an end is not even a doubt.

626. Nor does it mean anything to say: "The English name of this colour is *certainly* 'green',—unless, of course, I am making a slip of the tongue or am confused in some way."

627. Wouldn't one have to insert this clause into *all* language-games? (Which shows its senselessness.)

628. When we say "Certain propositions must be excluded from doubt", it sounds as if I ought to put these propositions—for example, that I am called L. W.—into a logic-book. For if it belongs to the description of a language-game, it belongs to logic. But that I am called L. W. does not belong to any such description. The language-game that operates with people's names can certainly exist even if I am mistaken about my name,—but it does presuppose that it is nonsensical to say that the majority of people are mistaken about their names.

629. On the other hand, however, it is right to say of myself "I cannot be mistaken about my name", and wrong if I say "perhaps I am mistaken". But that doesn't mean that it is meaningless for others to doubt what I declare to be certain.

630. It is simply the normal case, to be incapable of mistake about the designation of certain things in one's mother tongue.

631. "I can't be making a mistake about it" simply characterizes one kind of assertion.

632. Certain and uncertain memory. If certain memory were not in general more reliable than uncertain memory, i.e., if it were not confirmed by further verification more often than uncertain memory was, then the expression of certainty and uncertainty would not have its present function in language.

633. "I can't be making a mistake"—but what if I did make a mistake then, after all? For isn't that possible? But does that make the expression "I can't be etc." nonsense? Or would it be better to say instead "I can hardly be mistaken"? No; for that means something else.

634. "I can't be making a mistake; and if the worst comes to the worst I shall make my proposition into a norm."

635. "I can't be making a mistake; I was with him today."

636. "I can't be making a mistake; but if after all something *should* appear to speak against my proposition I shall stick to it, despite this appearance."

637. "I can't etc." shows my assertion of its place in the game. But it relates essentially to *me*, not to the game in general.
      If I am wrong in my assertion that doesn't detract from the usefulness of the language-game.

25.4.

638. "I can't be making a mistake" is an ordinary sentence, which serves to give the certainty-value of a statement. And only in its everyday use is it justified.

639. But what the devil use is it if—as everyone admits—I may be wrong about it, and therefore about the proposition it was supposed to support too?

640. Or shall I say: the sentence excludes a certain *kind* of failure?

641. "He told me about it today—I can't be making a mistake about that."—But what if it does turn out to be wrong?!—Mustn't one make a distinction between the ways in which something 'turns out wrong'?—How *can* it *be shewn* that my statement was wrong? Here evidence is facing evidence, and it must be *decided* which is to give way.

642. But suppose someone produced the scruple: what if I suddenly as it were woke up and said "Just think, I've been imagining I was called L. W.!"————well, who says that I don't wake up once again and call *this* an extraordinary fancy, and so on?

643. Admittedly one can imagine a case—and cases do exist—where after the 'awakening' one never has any more doubt which was imagination and which was reality. But such a case, or its possibility, doesn't discredit the proposition "I can't be wrong".

644. For otherwise, wouldn't all assertion be discredited in this way?

645. I can't be making a mistake,—but some day, rightly or wrongly, I may think I realize that I was not competent to judge.

646. Admittedly, if that always or often happened it would completely alter the character of the language-game.

647. There is a difference between a mistake for which, as it were, a place is prepared in the game, and a complete irregularity that happens as an exception.

648. I may also convince someone else that I 'can't be making a mistake'.

I say to someone "So-and-so was with me this morning and told me such-and-such". If this is astonishing he may ask me: "You can't be mistaken about it?" That may mean: "Did that really happen *this morning*?" or on the other hand: "Are you sure you understood him properly?" It is easy to see what details I should add to show that I was not wrong about the time, and similarly to show that I hadn't misunderstood the story. But all that can *not* show that I haven't dreamed the whole thing, or imagined it to myself in a dreamy way. Nor can it show that I haven't perhaps made some *slip of the tongue* throughout. (That sort of thing does happen.)

649. (I once said to someone—in English—that the shape of a certain branch was typical of the branch of an elm, which my companion denied. Then we came past some ashes, and I said "There, you see, here are the branches I was

speaking about". To which he replied "But that's an ash"—and I said "I always meant ash when I said elm".)

650. This surely means: the possibility of a *mistake* can be eliminated in certain (numerous) cases.—And one does eliminate mistakes in calculation in this way. For when a calculation has been checked over and over again one cannot then say "Its rightness is still only *very probable*—for an error may always still have slipped in". For suppose it did seem for once as if an error had been discovered—why shouldn't we suspect an error *here*?

651. I cannot be making a mistake about 12 × 12 being 144. And now one cannot contrast *mathematical* certainty with the relative uncertainty of empirical propositions. For the mathematical proposition has been obtained by a series of actions that are in no way different from the actions of the rest of our lives, and are in the same degree liable to forgetfulness, oversight and illusion.

652. Now can I prophesy that men will never throw over the present arithmetical propositions, never say that now at last they know how the matter stands? Yet would that justify a doubt on our part?

653. If the proposition 12 × 12 = 144 is exempt from doubt, then so too must non-mathematical propositions be.

26.4.51

654. But against this there are plenty of objections.—In the first place there is the fact that "12 × 12 etc." is a *mathematical* proposition, and from this one may infer that only mathematical propositions are in this situation. And if this

inference is not justified, then there ought to be a proposition that is just as certain, and deals with the process of this calculation, but isn't itself mathematical. I am thinking of such a proposition as: "The multiplication '12 × 12', when carried out by people who know how to calculate, will in the great majority of cases give the result '144'". Nobody will contest this proposition, and naturally it is not a mathematical one. But has it got the certainty of the mathematical proposition?

655. The mathematical proposition has, as it were officially, been given the stamp of incontestability. I.e.: "Dispute about other things; *this* is immovable—it is a hinge on which your dispute can turn."

656. And one can *not* say that of the proposition that *I* am called L. W. Nor of the proposition that such-and-such people have calculated such-and-such a problem correctly.

657. The propositions of mathematics might be said to be fossilized.—The proposition "I am called . . ." is not. But it too is regarded as *incontrovertible* by those who, like myself, have overwhelming evidence for it. And this not out of thoughtlessness. For, the evidence's being overwhelming consists precisely in the fact that we do not *need* to give way before any contrary evidence. And so we have here a buttress similar to the one that makes the propositions of mathematics incontrovertible.

658. The question "But mightn't you be in the grip of a delusion now and perhaps later find this out?"—might also be raised as an objection to any proposition of the multiplication tables.

659. "I cannot be making a mistake about the fact that I have just had lunch."

　　For if I say to someone "I have just eaten" he may believe that I am lying or have momentarily lost my wits but he won't believe that I am making a mistake. Indeed, the assumption that I might be making a mistake has no meaning here.

　　But that isn't true. I might, for example, have dropped off immediately after the meal without knowing it and have slept for an hour, and now believe I had just eaten.

　　But still, I distinguish here between different kinds of mistake.

660. I might ask: "*How* could I be making a mistake about my name being L. W.?" And I can say: I can't see how it would be possible.

661. How might I be mistaken in my assumption that I was never on the moon?

662. If I were to say "I have never been on the moon—but I may be mistaken", that would be idiotic.

　　For even the thought that I might have been transported there, by unknown means, in my sleep, *would not give me any right* to speak of a possible mistake here. I play the game *wrong* if I do.

663. I have a right to say "I can't be making a mistake about this" even if I am in error.

664. It makes a difference: whether one is learning in school what is right and wrong in mathematics, or whether I myself say that I cannot be making a mistake in a proposition.

665. In the latter case I am adding something special to what is generally laid down.

666. But how is it for example with anatomy (or a large part of it)? Isn't what it describes, too, exempt from all doubt?

667. Even if I came to a country where they believed that people were taken to the moon in dreams, I couldn't say to them: "I have never been to the moon.—Of course I may be mistaken". And to their question "Mayn't you be mistaken?" I should have to answer: No.

668. What practical consequences has it if I give a piece of information and add that I can't be making a mistake about it?

(I might also add instead: "I can no more be wrong about this than about my name's being L. W.")

The other person might doubt my statement nonetheless. But if he trusts me he will not only accept my information, he will also draw definite conclusions from my conviction, as to how I shall behave.

669. The sentence "I can't be making a mistake" is certainly used in practice. But we may question whether it is then to be taken in a perfectly rigorous sense, or is rather a kind of exaggeration which perhaps is used only with a view to persuasion.

27.4.

670. We might speak of fundamental principles of human enquiry.

671. I fly from here to a part of the world where the people have only indefinite information, or none at all, about the

possibility of flying. I tell them I have just flown there from. . . . They ask me if I might be mistaken.—They have obviously a false impression of how the thing happens. (If I were packed up in a box it would be possible for me to be mistaken about the way I had travelled.) If I simply tell them that I can't be mistaken, that won't perhaps convince them; but it will if I describe the actual procedure to them. Then they will certainly not bring the possibility of a *mistake* into the question. But for all that—even if they trust me—they might believe I had been dreaming or that *magic* had made me imagine it.

672. 'If I don't trust *this* evidence why should I trust any evidence?'

673. Is it not difficult to distinguish between the cases in which I cannot and those in which I can *hardly* be mistaken? Is it always clear to which kind a case belongs? I believe not.

674. There are, however, certain types of case in which I rightly say I cannot be making a mistake, and Moore has given a few examples of such cases.

I can enumerate various typical cases, but not give any common characteristic. (N. N. cannot be mistaken about his having flown from America to England a few days ago. Only if he is mad can he take anything else to be possible.)

675. If someone believes that he has flown from America to England in the last few days, then, I believe, he cannot be making a *mistake*.

And just the same if someone says that he is at this moment sitting at a table and writing.

676. "But even if in such cases I can't be mistaken, isn't it possible that I am drugged?" If I am and if the drug has taken away my consciousness, then I am not now really talking and thinking. I cannot seriously suppose that I am at this moment dreaming. Someone who, dreaming, says "I am dreaming", even if he speaks audibly in doing so, is no more right than if he said in his dream "it is raining", while it was in fact raining. Even if his dream were actually connected with the noise of the rain.

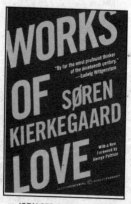